CONTENTS

INTRODUCTION

The Megaloptera and Neuroptera (or Planipennia) are usually regarded as two closely related orders of the Insecta and some specialists combine them into one order of Neuroptera. Both orders belong to the Endopterygota (or Holometabola), i.e. those insects with a complete metamorphosis internally in the larval and pupal stages, with a larva that differs from the adult in its structure and habits, and with a life cycle divided into four definite stages of egg, larva, pupa and imago. The name Neuroptera (Greek *neuron*=nerve; *pteron*=wing) refers to the nerve-like network of veins in the wings. Linnaeus originally used this name for all insects with delicate, net-veined wings and the Linnaean order included the exopterygote Ephemeroptera and Odonata. This miscellany of veined-winged insects was gradually split into several orders and now the order Neuroptera includes a relatively small number of species.

The Megaloptera and Neuroptera are among the most primitive of the endopterygote insects, being close to the ancestral stock of the Panorpoid complex which also includes the Mecoptera, Lepidoptera, Trichoptera, Siphonaptera and Diptera. Their characteristic wings are very similar to those of Palaeozoic fossil insects and the first fossil records of the two orders are from the Lower Permian of Kansas. About 4500 species are known at present and sixty species have been recorded in Britain.

GENERAL CHARACTERS

Adults of both orders are usually soft-bodied with long slender antennae, with no abdominal cerci, with their tarsi having five segments (the tarsus is the distal part of the leg and usually ends in a pair of claws), and with their mouthparts adapted for biting. They have two pairs of very similar membraneous wings which are held at rest like a peaked roof or tent over the abdomen. The hindwings are only slightly smaller than the forewings and are of similar shape. Although the wing venation is generally regarded as primitive, there are usually many accessory veins and numerous costal veinlets giving a ladder-like effect along the anterior edge of the wing. The radial sector in the forewing (third longitudinal vein behind the anterior edge of the wing) emits two or more branches. Adult Neuroptera are usually crepuscular but adult Megaloptera fly readily during warm sunny weather. Species with aquatic larvae do not lay their eggs in water but deposit them on objects overhanging the water.

Pupae of both orders are exarate, i.e. the appendages are not adherent to the rest of the body and can be used in locomotion. Species with aquatic larvae pupate on land but always close to water.

Larvae of both orders have a well-sclerotized head with obvious forwardly-directed mouthparts and large 'jaws' (mandibles). Their thoracic legs are usually long and there are no abdominal prolegs. The whole body or at least the abdomen is more or less elongately fusiform in shape. Aquatic larvae have seven pairs of lateral or ventral abdominal gills. All larvae are active predaceous insects, with biting mouthparts in the Megaloptera and sucking mouthparts in the Neuroptera.

CLASSIFICATION AND CHECK LIST

The British species of Megaloptera fall naturally into the suborders Raphidioidea and Sialoidea, and each suborder has only one family, namely the Raphidiidae with four species in the genus *Raphidia*, and the Sialidae with two species in the genus *Sialis*. Larvae of *Raphidia* are terrestrial and the adults are known as snake-flies because of their narrow elongate prothorax which forms a kind of neck on which the head is held like that of a snake ready to strike. Larvae of *Sialis* are aquatic and the adults are known as alder-flies, no doubt because they are often found on alders that overhang the water.

The British species of Neuroptera fall into five families, namely the Coniopterygidae with seven species in five genera, the Osmylidae with only one species, the Sisyridae with three species in one genus, the Hemerobiidae with twenty-nine species in eight genera and the Chrysopidae with fourteen species in two genera. Adults of the two largest families are known as the brown (Hemerobiidae) and green (Chrysopidae) lacewings, and their larvae are always terrestrial. Larvae of the Coniopterygidae are also terrestrial and the minute adults are the smallest and most aberrant of the Neuroptera with their opaque white, waxy wings which have a greatly reduced venation. Larvae of the

Sisyridae are aquatic and the adults are sometimes known as sponge-flies because the larvae are usually found on freshwater sponges on which they feed. The Osmylidae are closely allied to the Sisyridae and larvae of the single British species are semi-aquatic, living under stones or in moss at the edges of streams.

Adults of the British species of Megaloptera and Neuroptera were described by McLachlan (1868) and Killington (1929), and keys to adults are included in the handbook of Fraser (1959). Killington (1936, 1937) has also written a monograph of the British Neuroptera. Only two of the six species of Megaloptera and four of the fifty-four species of Neuroptera are aquatic or semi-aquatic in the larval stage. These six species* are:

> MEGALOPTERA
> Family SIALIDAE
> Genus SIALIS Latreille, 1803
> S. lutaria (Linnaeus, 1758)
> (=flavilatera Linnaeus, 1758)
> S. fuliginosa Pictet, 1836

> NEUROPTERA
> Family OSMYLIDAE
> Genus Osmylus Latreille, 1802
> O. fulvicephalus (Scopoli, 1763)
> (=chrysops Linnaeus, 1758)

> Family SISYRIDAE
> Genus SISYRA Burmeister, 1839
> S. fuscata (Fabricius, 1793)
> S. dalii McLachlan, 1866
> S. terminalis Curtis, 1854

COLLECTION AND PRESERVATION

Adults are most readily collected by beating bushes and trees, or sweeping low herbage near ponds, lakes and streams. Osmylus is sometimes found on the underside of low bridges over woodland streams. Some kind of bottom sampler is required to collect larvae of Sialis from the mud of ponds, lakes and streams. Larvae may occasionally be found under large stones in streams. Sisyra larvae are found in

* See footnote on p. 14.

freshwater sponges of lakes, canals and streams. *Osmylus* larvae live in the moss bordering woodland streams.

If a pinned collection is desired, adults should be brought back alive and killed with ether or chloroform. Stainless steel pins and flat setting boards should be used, and the specimen set in the same way as Lepidoptera, the wings being held in place by strips of paper or cellophane during drying. They should be set as soon as possible after killing, since *Sisyra*, in particular, dries very rapidly and becomes brittle.

Adults, pupae and larvae may also be preserved in fluid, e.g. 70% alcohol, dilute formaldehyde (one part of 40% formaldehyde to nineteen parts of water), or a mixture of alcohol and dilute formaldehyde. A suitable fixative fluid is Pämpel's mixture (four parts of glacial acetic acid, thirty parts of distilled water, six parts of 40% formaldehyde and fifteen parts of 95% alcohol). An excellent killing fluid and preservative for material suitable for dissection is K.A.A.D. (one part of kerosene, ten parts of 95% alcohol, two parts of glacial acetic acid, one part of dioxan).

KEY TO ADULTS

The general characters given in the introduction separate the Megaloptera and all families of Neuroptera, except the Coniopterygidae, from all other orders of insects.   The most useful character is the two pairs of membraneous wings covered with a delicate network of veins, including many cross-veins.   Some species of Lepidoptera, Trichoptera, Ephemeroptera, Plecoptera, Odonata and Mecoptera have similar wings but can be distinguished from the Megaloptera and Neuroptera by the following characters.   In the Lepidoptera (butterflies and moths), the wings are covered with minute scales and there is usually a coiled proboscis (tongue) on the underside of the head.   In the Trichoptera (caddis-flies), the wings are usually densely covered with hairs, the hind-wing is usually broader than the forewing and there are relatively few cross-veins.   In the Ephemeroptera (mayflies), the hindwings, when present, are always considerably smaller than the forewings and there are two or three abdominal cerci.   In the Odonata (dragonflies and damsel-flies), the antennae are always very short.   In the Plecoptera (stoneflies), abdominal cerci are clearly seen in most species, the hindwings are usually considerably larger than the forewings and all tarsi have three segments.   The Mecoptera (scorpion-flies) are closely related to the Megaloptera and Neuroptera, but are easily distinguished from them by the shape of the head, which is extended downwards to form a beak or rostrum.

As the adults of species with terrestrial larvae are often found near water and may be confused with adults of species with aquatic larvae, a general key to families is given before the key to species of the three families with aquatic or semi-aquatic larvae.   The Coniopterygidae are not included in the key to families because they differ considerably from other species of Neuroptera, being minute and covered with white waxy powder.   Their wing venation is greatly reduced, with few cross-veins and usually with the hindwings much smaller than the forewings.

KEY TO FAMILIES

1 Branches of the wing-veins are not frequently bifurcated at the wing margin (Fig. 2A, B). Third or fourth segment of tarsus is bilobed (Fig. 1A)—   MEGALOPTERA 2

— Branches of the wing-veins are usually conspicuously bifurcated at the margins of the wings (Fig. 2C, D, E, F). All tarsal segments are similar and narrow (Fig. 1B)—   NEUROPTERA 3

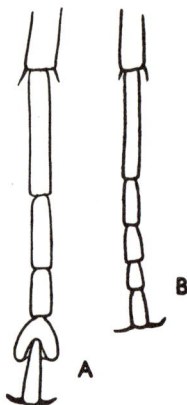

Fig. 1.   Tarsi of: A, *Sialis lutaria*; B, *Sisyra fuscata*.

2 Anterior part of thorax (prothorax) is elongated to form a kind of neck on which the head is held like that of a snake ready to strike. Wings with a pterostigma (*Pt* in Fig. 2A) which is an opaque chitinous cell. Third segment of tarsus is bilobed—

RAPHIDIIDAE (snake-flies).
Larvae terrestrial.

— Prothorax is short and broad. Wings without a pterostigma (Fig. 2B). Fourth segment of tarsus is bilobed (Fig. 1A). Wings at rest form a broad 'roof' (Fig. 3A)—   SIALIDAE (alder-flies).
Larvae aquatic.

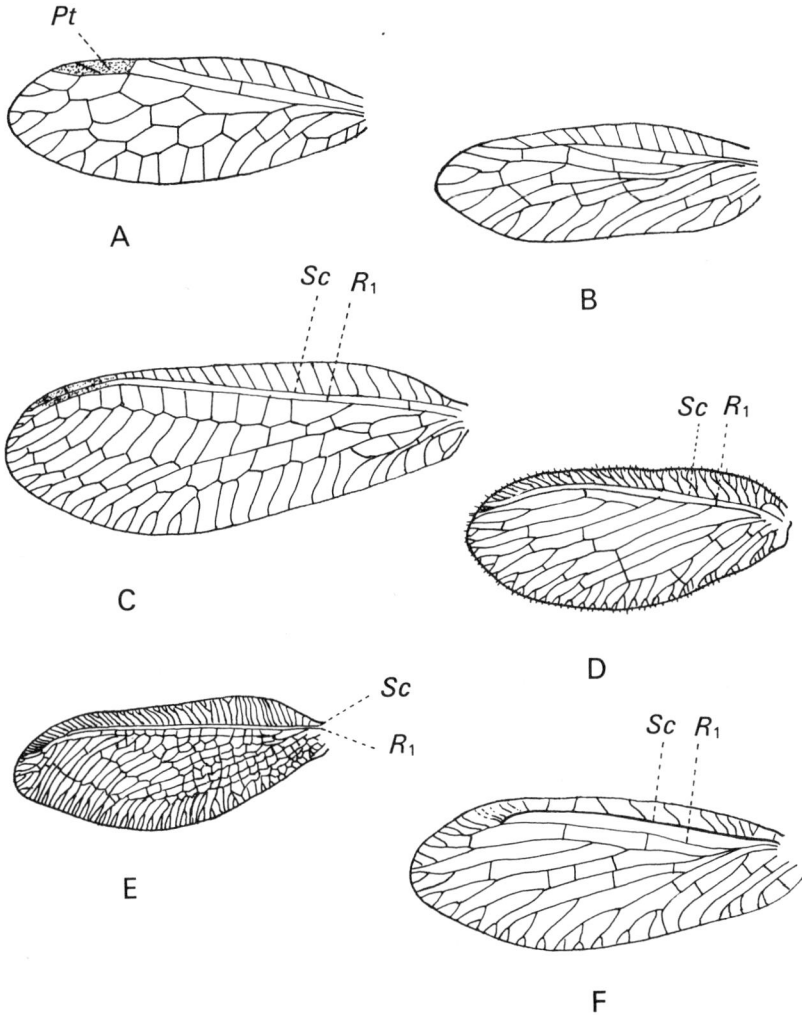

Fig. 2. Forewings of: A, Raphidiidae (length of wing 10–14 mm); B, Sialidae (8–17 mm); C, Chrysopidae (10–25 mm); D, Hemerobiidae (3–16 mm); E, Osmylidae (19–23 mm); F, Sisyridae (5–6 mm).

3(1) Subcosta (Sc) and radius (R_1) do not join near the apex of the
forewing (beware of cross-veins) (Fig. 2C, D) and run close together
near the anterior edge of the wing—   **4**

— Subcosta (Sc) and radius (R_1) fuse near the apex of the forewing
(Fig. 2E), or the subcosta bends posteriorly to meet the radius near
the apex of the forewing (Fig. 2F)—   **5**

4 Large to medium-sized insects coloured yellow or bright green, with
golden eyes and filiform (thread-like) antennae. There are only a
few longitudinal veins in the wings (Fig. 2C) and hairs are limited to
the veins, cross-veins and edge of the wing—
                              CHRYSOPIDAE (green lacewings).
     Larvae terrestrial.

— Small grey or brown insects with moniliform (like a string of beads)
antennae.   There are several longitudinal veins in the wings (Fig.
2D) which are covered with short fine hairs on both the veins and the
membrane.   Minute wart-like tubercles (trichosors), each with two
or three bristles, are situated along the edge of the wing between the
terminal branches of the veins—
                              HEMEROBIIDAE (brown lacewings).
     Larvae terrestrial.

5(3) Wings are large (forewing 19-23 mm) and transparent, with
numerous cross-veins.   Subcosta (Sc) and radius (R_1) run very close
together and fuse near the apex of the wing (Fig. 2E).   Wing
membrane has darkish brown spots or blotches (Fig. 3C)—
                                                    OSMYLIDAE.
     Larvae semi-aquatic.

— Wings are small (forewing 5-6 mm) with few cross-veins.   Subcosta
(Sc) bends to meet the radius (R_1) near the apex of the forewing
(Figs 2F, 3B)—   SISYRIDAE (sponge-flies).
     Larvae aquatic.

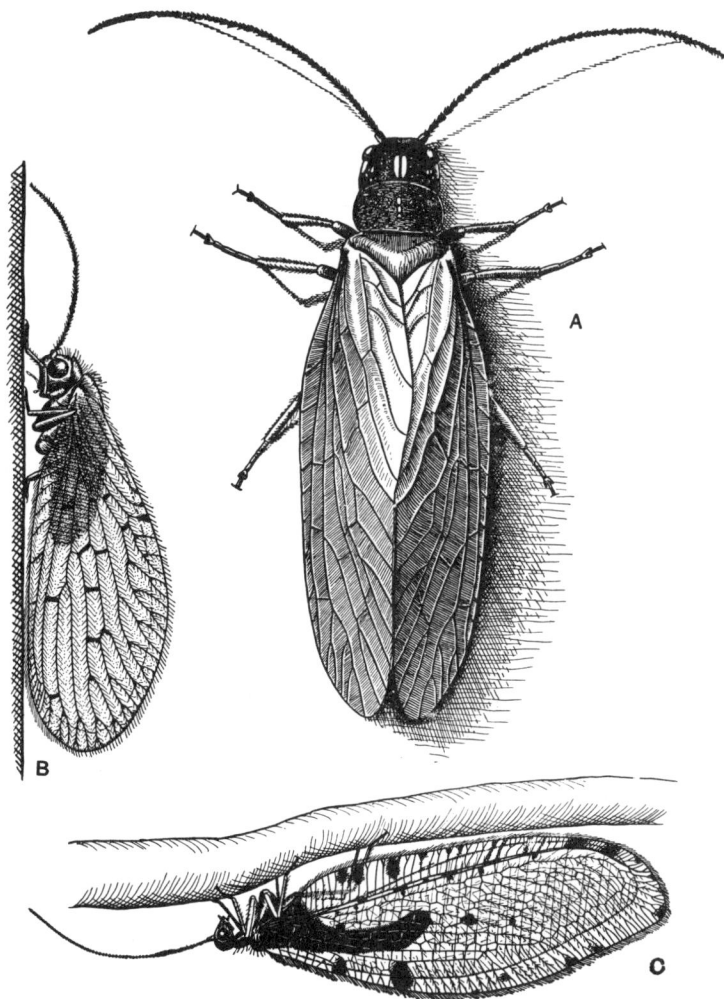

Fig. 3. Megaloptera and Neuroptera in resting positions: A, *Sialis lutaria* × 6·4; B, *Sisyra dalii* × 9·5; C, *Osmylus fulvicephalus* × 2·8.

KEY TO SPECIES WITH AQUATIC OR SEMI-AQUATIC LARVAE

Family SIALIDAE.   Genus SIALIS Latreille.

Although the wing venation is frequently used to separate the two species* of *Sialis* (e.g. Fraser 1959, Kimmins 1962), it is rather variable and is therefore not a reliable character (Seitz 1940, Hoffmann 1962, Selman 1965, Vaňhara 1970).   As the characters of the external genitalia are much more constant, they are more satisfactory for an unequivocal determination (Kaiser 1950, Meinander 1962, Vaňhara 1970).

WING VENATION

1   Cross-vein between subcosta and radius in forewing (*Sc-R* in Fig. 4A) usually meets subcosta between the fourth and fifth costal cross-veins, but occasionally it is opposite to the fifth costal cross-vein or it meets subcosta between the fifth and sixth costal cross-veins. There are nine to thirteen costal cross-veins.   Three gradate cross-veins (*Gr* in Fig. 4A) lie almost in a straight line (anterior gradate cross-vein is often faint).   Length of forewing 8-11 mm in male, 11-15 mm in female.   Flight period late April-June, occasionally July.   Adults are usually found near the banks of ponds, lakes and sluggish streams and rivers where there is an abundance of silt—
                                    Sialis lutaria (Linn.)

—   Cross-vein between subcosta and radius in forewing (*Sc-R* in Fig. 4B) usually meets subcosta between seventh and eighth costal cross-veins, but occasionally it is opposite to the sixth costal cross-vein or it meets subcosta between the sixth and seventh costal cross-veins.   There are fourteen to sixteen costal cross-veins.   Three gradate cross-veins (*Gr* in Fig. 4B) do not lie in a straight line. Length of forewing 9-13 mm in male, 13-17 mm in female.   Flight period May-June.   Adults are usually found near the banks of streams and rivers—   **Sialis fuliginosa** Pict.

* Whilst this publication was at the printers, two adults of a third species, *Sialis nigripes* Pict., were discovered in recent (1976) material from Ireland, and a re-examination of material in the British Museum (Natural History) has shown that four specimens have been wrongly identified and are actually *S. nigripes* (P. C. Barnard, personal communication). One of these adults was collected in Ireland in 1914 and the others were caught in the English counties of Surrey in 1867, Dorset in 1951 and Somerset in 1955.   It would appear from the small number of records that this species occurs in rivers in southern England and Ireland.   Both Kaiser (1950) and Vaňhara (1970) provide characters for separating adults of this species from *S. lutaria and S. fuliginosa*, and Kaiser (1950, 1961) has described the life cycle in Denmark.

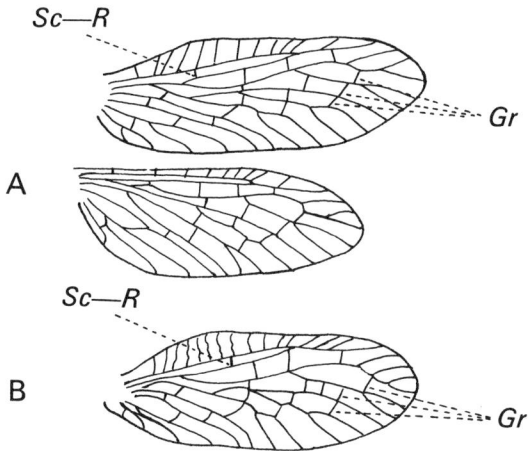

Fig. 4. A, wings of *Sialis lutaria* (length of forewing 8–15 mm); B, forewing of *S. fuliginosa* (length of wing 9–17 mm); *Sc–R*, cross-vein between subcosta and radius; *Gr*, gradate cross-veins.

1   Ninth tergite is almost as long as ninth sternite (9*t* and 9*st* in Fig. 5A). Ninth sternite forms a subgenital plate which is spoon-shaped and nearly as long as broad (9*st* in Fig. 5A, B), overlapping the end of the abdomen when closed. Membraneous, tubular utriculi are well developed (*u* in Fig. 5B)—   **Sialis lutaria**

—   Ninth tergite is much longer than ninth sternite (9*t* and 9*st* in Fig. 5C). Ninth sternite forms a short subgenital plate which is distinctly broader than long (9*st* in Fig. 5D) and never overlaps the end of the abdomen. Utriculi are absent. Tenth tergite (anal segment) is distinctly beak-like in profile (Fig. 5C) and terminates in two projections (Fig. 5D)—   **Sialis fuliginosa**

1   Eighth sternite does not narrow markedly in the mid-line and has a medial tooth on the posterior margin which is slightly concave (8*st* in Fig. 5E). Seventh sternite usually has rounded posterior corners—   **Sialis lutaria**

—   Eighth ·sternite narrows markedly in the mid-line and is concave towards the anterior with no medial tooth on the posterior margin (8*st* in Fig. 5F). Seventh sternite usually has angular posterior corners—   **Sialis fuliginosa**

Family OSMYLIDAE.   Genus Osmylus Latreille

There is only one British species (Fig. 3C). Length of forewing 19-23 mm. Flight period May-August. Adults are usually found near small woodland streams that have moss along their banks and thus provide a suitable habitat for the larvae—

Osmylus fulvicephalus (Scop.)

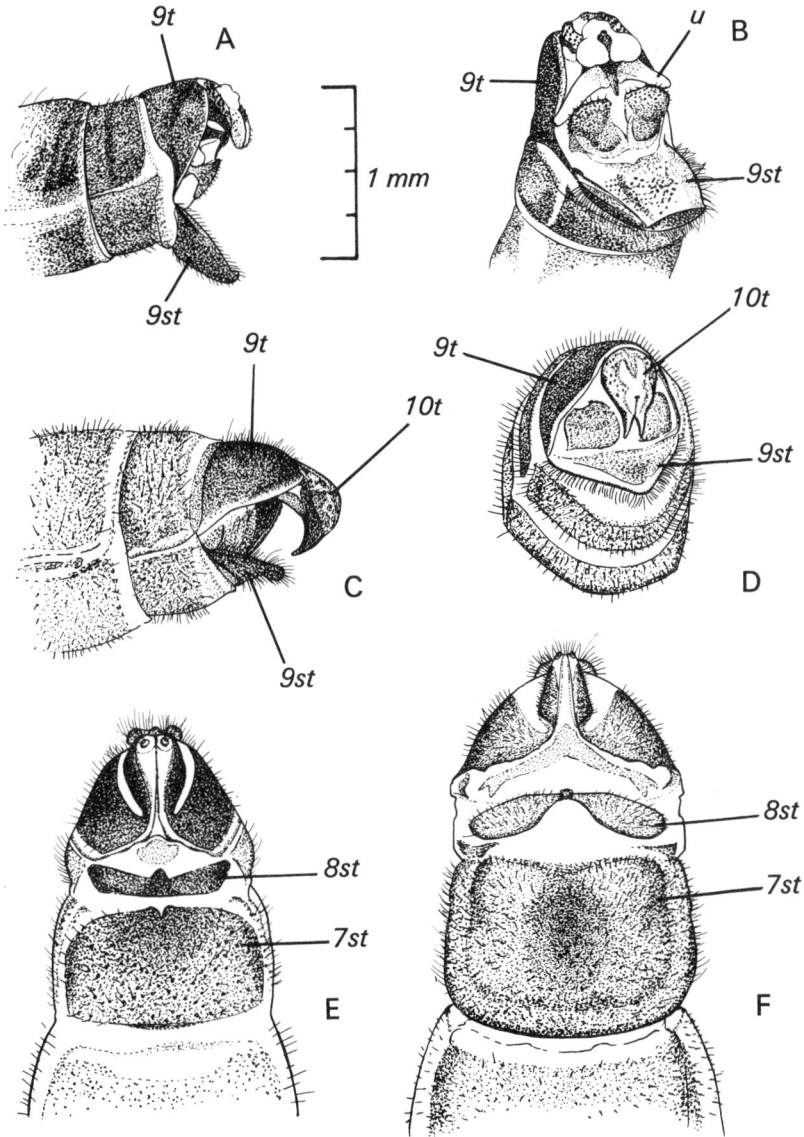

Fig. 5. External genitalia of *Sialis*: A, male *S. lutaria* in lateral view; B, male *S. lutaria* from posterior; C, male *S. fuliginosa* in lateral view; D, male *S. fuliginosa* from posterior; E, female *S. lutaria* in ventral view; F, female *S. fuliginosa* in ventral view. *9t*, ninth tergite; *10t*, tenth tergite; *9st*, ninth sternite; *8st*, eighth sternite; *7st*, seventh sternite; *u*, utriculus.

Family SISYRIDAE.   Genus SISYRA Burmeister

Although the general characters are usually satisfactory for separating the three species of *Sisyra*, they are sometimes obscure in specimens preserved in fluid and are best seen in dry specimens mounted on pins. Therefore the external genitalia provide the most consistent characters for separating the adults.

GENERAL CHARACTERS

1   Antennae are dark brown or black with pale yellow or yellowish-white tips.   Wings are grey, almost colourless, with pale cross-veins and usually with only one cross-vein in the apical third of the wing, placed between R_1 and R_2 (Fig. 6A).   Length of forewing 5-6 mm. Flight period May-July.   Adults are usually found near the banks of ponds, canals, rivers and slow streams, containing sponges on which the larvae live and feed—   **Sisyra terminalis** Curt.

—   Antennae are dark throughout.   Wings are a dingy-brown, blackish-brown or reddish-brown—   **2**

2   The gradate cross-veins are not darker than the rest of the wings. Only one cross-vein is usually present in the apical third of the wing and this is usually placed between R_1 and R_2 (Fig. 6B), or there are no cross-veins in the apical region.   Length of forewing 5-6 mm. Flight period May-September.   Adults are usually found near the banks of lakes, canals and streams which must contain sponges—
Sisyra fuscata (Fabr.)

—   The gradate cross-veins of the forewing and the membrane next to the cross-veins are distinctly darker than the rest of the wing, and the cross-veins appear to be thicker than the other wing-veins (Figs 3B, 6C).   One, two or three cross-veins are present in the apical third of the wing (Fig. 6C).   Length of forewing 5-6 mm. Flight period June-August.   Little is known about the ecology of this rare species but adults apparently occur in the same places as *S. fuscata*—   **Sisyra dalii** McLach.

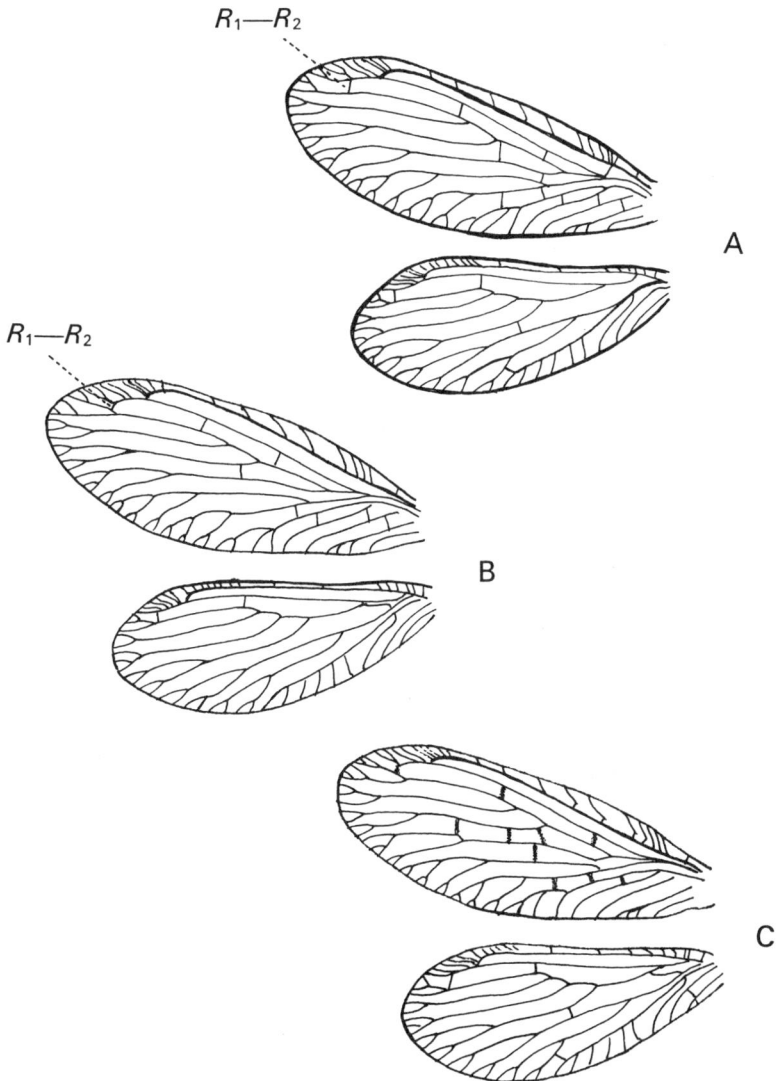

Fig. 6.   Wings of *Sisyra* (length of forewing 5–6 mm): A, *S. terminalis*; B, *S. fuscata*; C, *S. dalii*.

MALE GENITALIA

1 Claspers are inwardly curved and taper to an acute apex (Fig.
   7A, B)—   **Sisyra fuscata**

— Claspers are slightly convergent and truncate at the apex (Fig.
   7C, D)—   **2**

2 Claspers have sinuate margins; the lower margin ends in a small
   acute process which curves upwards and inwards, whilst the upper
   margin ends in three processes, a short stout spine, a long stout spine
   and a small tooth-like process (Fig. 7C)—   **Sisyra terminalis**

— The upper margin of each clasper ends in a short inwardly-curved
   tooth below which are one or two blunt projections, with a slightly
   longer acute process near the lower margin (Fig. 7D)—
                                                    Sisyra dalii

FEMALE GENITALIA

Although there are slight differences in the genitalia, especially in the
form of the valve-like processes of the ninth sternite, these differences
are not considered large enough to provide good characters for separating
females of the three species (Fig. 7E, F, G). Figures are provided so
that males and females can be recognised.

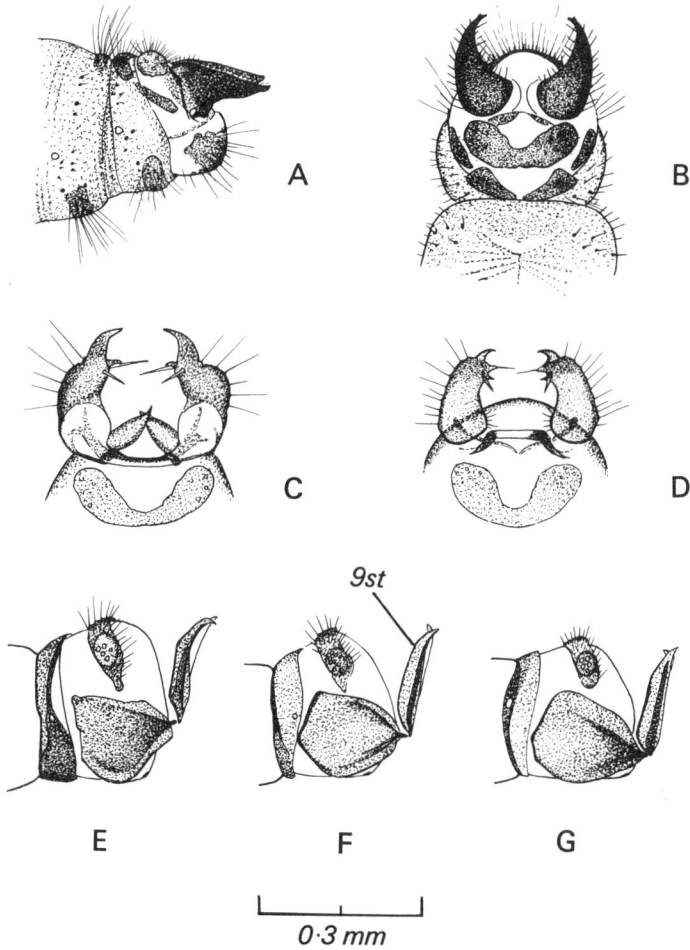

Fig. 7. External genitalia of *Sisyra*: A, male *S. fuscata* in lateral view; B, male *S. fuscata* in dorsal view; C, male *S. terminalis* in dorsal view; D, male *S. dalii* in dorsal view; E, female *S. fuscata* in lateral view; F, female *S. terminalis* in lateral view; G. female *S. dalii* in lateral view. *9st*, ninth sternite.

KEY TO LARVAE

The general characters given in the introduction separate the Megaloptera and Neuroptera from all other orders of insects. The most useful characters are the fusiform shape of the body and the large 'jaws' which are either large mandibles with sharp teeth for biting (Megaloptera), or modified mandibles and maxillae which fit together to form two forwardly-directed 'spears' or 'jaws' used for sucking the contents of the prey (Neuroptera). Some beetle larvae may be mistaken for larvae of Neuroptera, but they usually have a parallel-sided body and smaller jaws. Larvae of Gyrinidae (whirligig beetles) may sometimes be mistaken for *Sialis* larvae because they also have lateral abdominal gills. However, the gills are jointed in *Sialis* and the abdomen ends in a long tail, whilst the gills of Gyrinidae are unjointed and the abdomen ends in four hooks instead of a tail. Larvae of Trichoptera (caddis-larvae) may sometimes be mistaken for larvae of *Sialis* or *Sisyra* because some free-living and net-spinning species also have abdominal gills and prominent 'jaws'. However, the gills are never jointed in Trichoptera and the abdomen ends in a pair of pro-legs, each carrying a curved claw.

As terrestrial larvae of Megaloptera and Neuroptera are often found near water and sometimes fall into the water, they may be confused with the truly aquatic larvae of *Sialis* and *Sisyra*, and especially with the semi-aquatic larvae of *Osmylus fulvicephalus*. Therefore a key to families is given before the key to aquatic and semi-aquatic larvae.

KEY TO FAMILIES

1   Larvae have powerful mandibles armed with teeth and used for biting (Fig. 8A)—   MEGALOPTERA 2

—   Larvae have apposed mandibles and maxillae which are hollowed out to form two tubes or suctorial jaws projected in front of the head (Fig. 8B)—   NEUROPTERA 3

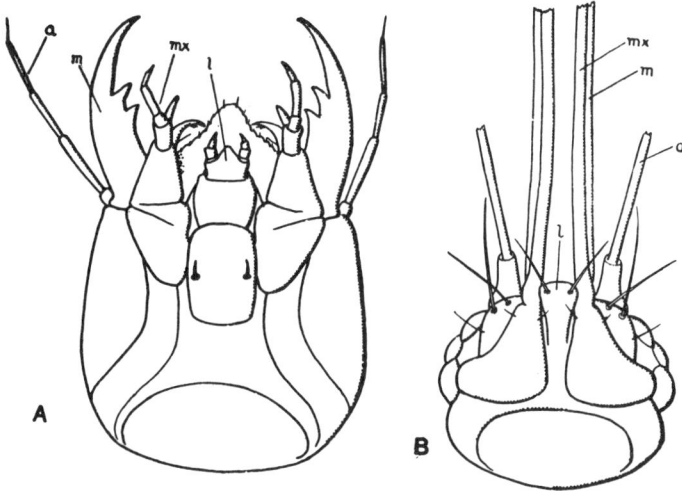

Fig. 8.   Ventral view of larval heads of: A, *Sialis lutaria*; B, *Sisyra* sp. *a*, antenna, *m*, mandible, *mx*, maxilla, *l*, labium.
(Only the basal part of the antennae, mandibles and maxillae are shown in *Sisyra*.)

2   Larvae are terrestrial and have no abdominal processes or appendages—   RAPHIDIIDAE

—   Larvae are aquatic and have a long tapering tail and seven pairs of segmentally-arranged jointed gills on the lateral margins of the abdomen (Fig. 9)—   SIALIDAE

3(1) Larvae are aquatic and live in or on sponges (Fig. 10B).   There are seven pairs of segmentally-arranged jointed gills on the ventral surface of the abdomen (Fig. 11).   The mouthparts are modified to form long slender sucking tubes (Fig. 8B) and each tarsus has only one claw, with no empodium next to the claw—   SISYRIDAE

—   Larvae are terrestrial or live at the margins of streams, and have no gills.   Each tarsus has two claws and an empodium between the claws (the empodium is a cushion-like structure which may be a tactile organ, or may be used to hold on to smooth surfaces)—   4

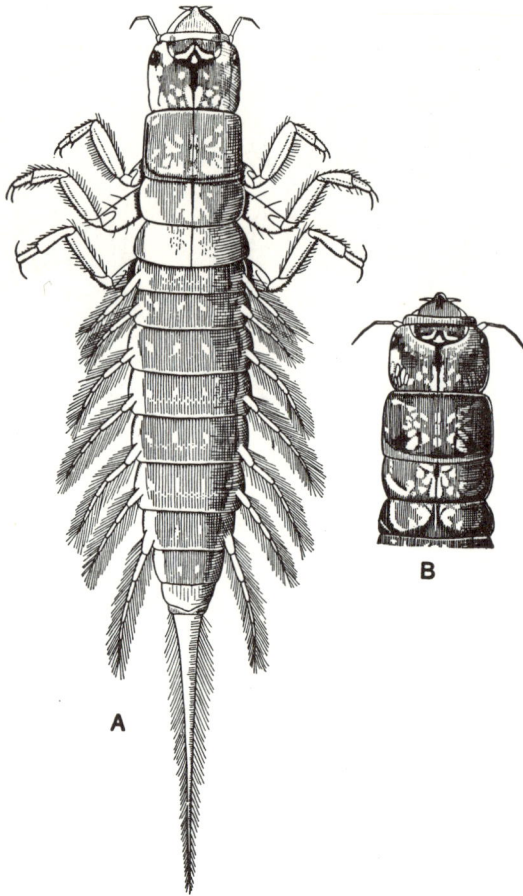

Fig. 9.   A, larva of *Sialis fuliginosa*; B, head and thorax of larva of *S. lutaria* × 4·5.

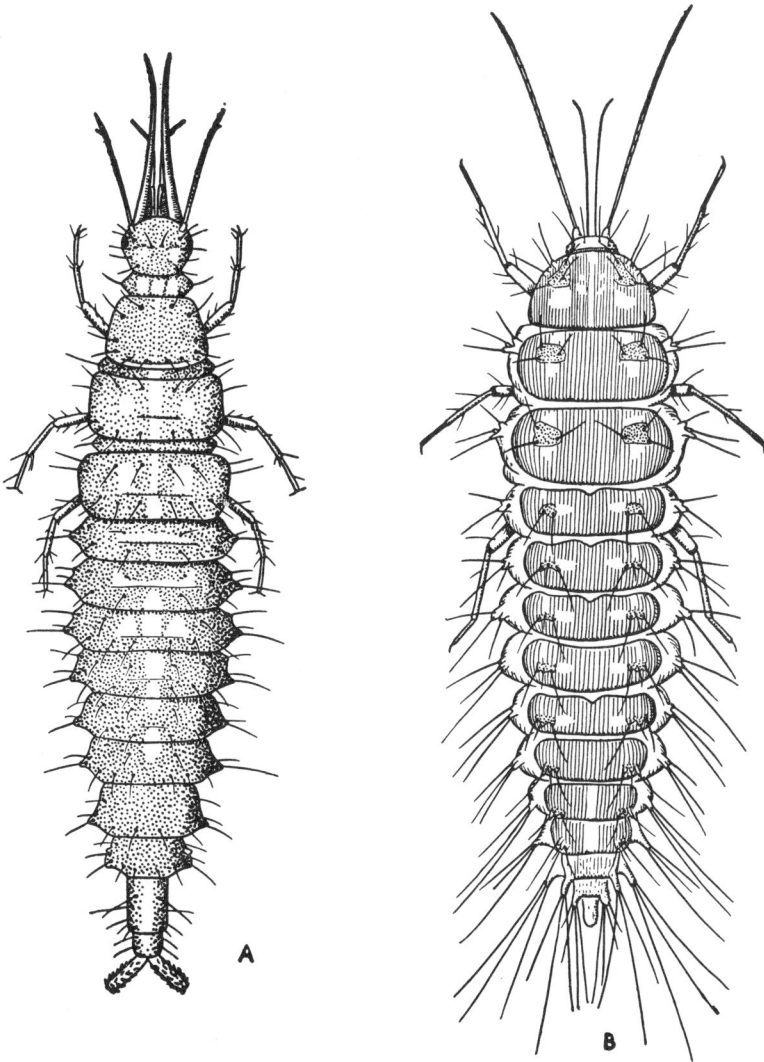

Fig. 10. Larvae of: A, *Osmylus fulvicephalus* after Killington, × 8; B, *Sisyra* sp. × 18.

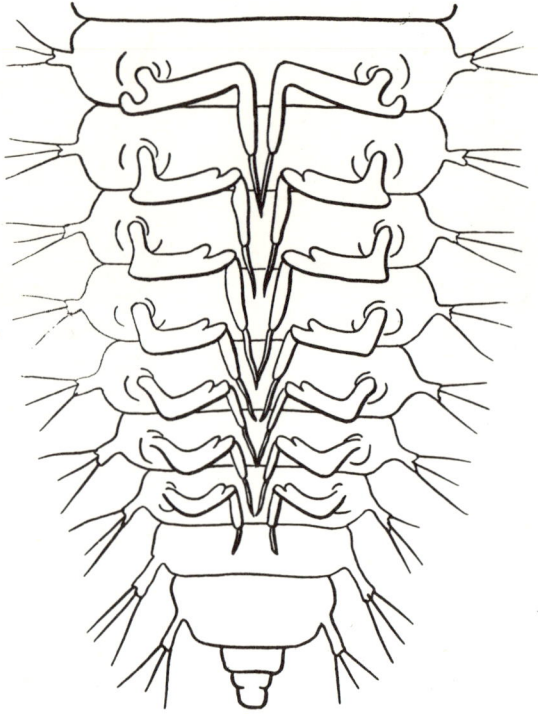

Fig. 11. Ventral view of abdominal segments and gills of larva of *Sisyra fuscata* adapted from Lestage.

4   Larvae are semi-aquatic, living in moss at the edge of streams. Mandibles and maxillae form long slender stylet-like tubes which project in front of the head and are slightly curved upwards (Fig. 10A).   The body bears numerous short setae and the abdomen terminates in a pair of eversible finger-like processes which are covered with recurved hooks. Empodium is tapered with short fine hairs at the apex—   OSMYLIDAE

—   Larvae are terrestrial.   Jaws are prominent caliper-like structures that curve inwards.   The abdomen terminates in a pair of eversible pads, not covered with hooks—   **5**

5 Jaws are long, slender and strongly curved. Empodium is trumpet-shaped. Body is usually clothed with long setae and there are usually prominent dorsolateral tubercles, bearing long setae—
CHRYSOPIDAE

— Jaws are stout and only slightly curved. Empodium is short and pad-like. Body is rather sparingly clothed with fine and inconspicuous hairs and there are no prominent tubercles—
HEMEROBIIDAE

KEY TO SPECIES WITH AQUATIC OR SEMI-AQUATIC LARVAE

Family SIALIDAE. Genus SIALIS Latreille

Only one key to larvae has been found in the literature and that is in the previous FBA publication by Kimmins (1962). He used the following characters to separate the two species*: abdomen purplish-pink, head and thorax dark brown in *S. lutaria*; abdomen orange-yellow, head and thorax yellow-brown in *S. fuliginosa*. Unfortunately, the colour of the larvae is very variable and therefore these characters are unreliable. The four characters used in the following key are described for the first time and are the only morphological features that were found to be consistently different in the two species. They are applicable to instars four to ten, but may not always work for instars two and three, and are definitely not applicable to the first instar, which is unlike all the other instars. Although the four characters have not been described before, E. W. Kaiser (personal communication) has also found that the same characters can be used to separate larvae of *Sialis* in Denmark.

The form of the mandibles, maxillae, labium, gular sclerite, abdominal gills and terminal process is very similar in the two species, but there are some differences in the number of spines and hairs on the legs. For example, the number of spines on the anterior face of the posterior tibia is usually between six and seven in *S. lutaria* and between three and five in *S. fuliginosa*, but exactly the opposite is seen in some specimens. There are similar differences between the anterior legs, but none of these differences is consistent.

*A third species, *S. nigripes*, has now been recorded in Britain, (see footnote on p. 14) and the author would like to see any larvae that do not agree with this key.

1   Postoccipital sutures are distinctly arched and end at a short distance from the epicranial suture (Fig. 12A). Anterior edge of the labrum is strongly crinkled (Fig. 12C). A distinct dark band usually runs from the eyes in an oblique direction towards the posterior. Each abdominal tergum usually has a large median pale spot and two small lateral pale spots (Fig. 12E), but these spots may be fused in some specimens (Fig. 12F) or difficult to see in very dark specimens. As the life cycle takes at least two years, larvae are found throughout the year in ponds, lakes and sluggish streams and rivers where there is an abundance of silt. Length of larva, without terminal filament, is about 1 mm in the first instar and between 12 and 20 mm in the tenth and final larval instar—   **Sialis lutaria** (Linn.)

—   Postoccipital sutures are not arched and meet the epicranial suture directly or in the form of a groove (Fig. 12B). Anterior edge of the labrum is usually smooth (apart from the stout spines) but may appear slightly crinkled in pale specimens (Fig. 12D). There is no dark band near the eyes. Each abdominal tergum has two pale lateral spots with no median spot (Figs 9A, 12G), but these spots may be difficult to see in pale specimens. The life cycle probably takes two years and larvae are found throughout the year in moderately fast streams and rivers. Length of larva, without terminal filament, is about 1 mm in the first instar and between 16 and 26 mm in the tenth instar—   **Sialis fuliginosa** Pict.

Family OSMYLIDAE.   Genus OSMYLUS Latreille

There is only one British species (Fig. 10A). The life cycle takes about one year and larvae can be found throughout the year in the moss bordering woodland streams. Larvae may be difficult to find from late September to March because they overwinter deep in the roots of the moss and are in a state of hibernation known as diapause. Length of larva, including jaws, is about 5 mm in the first instar, about 9 mm in the second instar and about 15 mm in the third and final larval instar—

Osmylus fulvicephalus (Scop.)

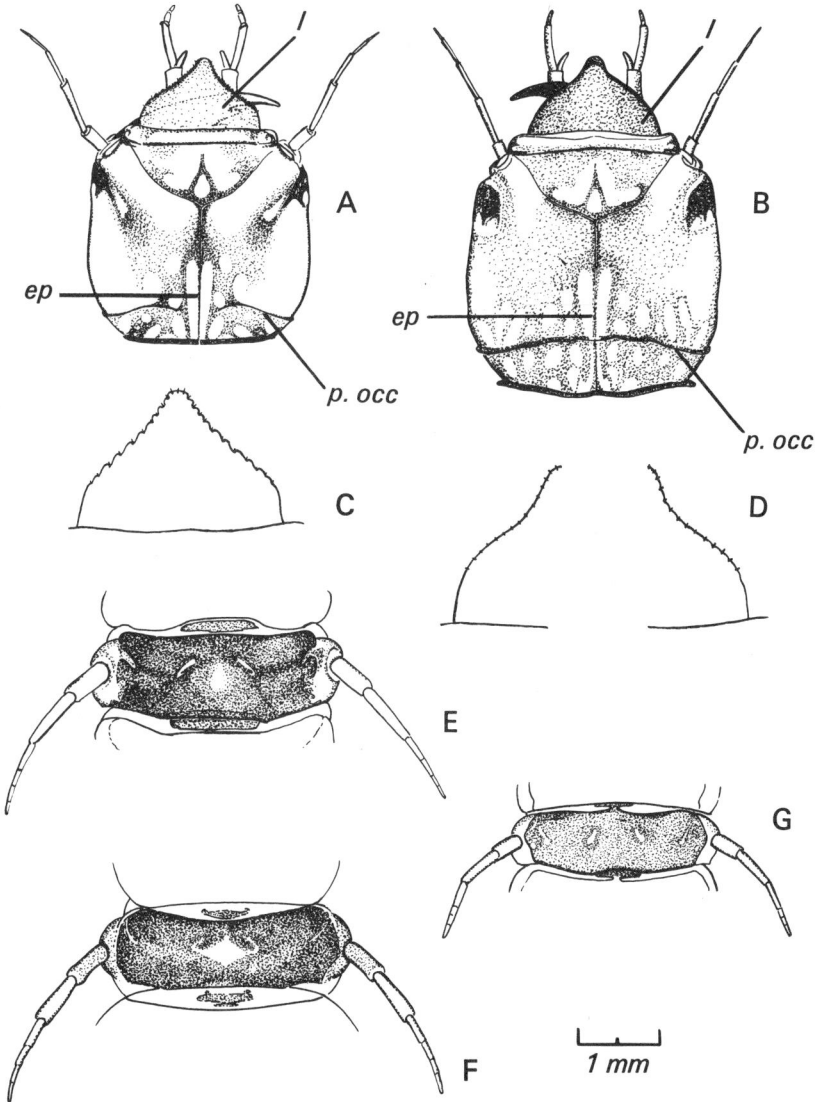

Fig. 12. Larvae of *Sialis*: A, head of *S. lutaria* in dorsal view; B, head of *S. fuliginosa* in dorsal view; C, labrum of *S. lutaria* in dorsal view; D, labrum of *S. fuliginosa* in dorsal view, showing smooth edge of most specimens and slightly crinkled edge of pale specimens; E, F, second abdominal segment of *S. lutaria* in dorsal view, showing typical pattern and fused pattern; G, second abdominal segment of *S. fuliginosa* in dorsal view.   *l*, labrum; *ep*, epicranial suture; *p. occ*, postoccipital suture.

Family SISYRIDAE. Genus SISYRA Burmeister

It has not been possible to construct a key to larvae because of the lack of specimens. The larva of *S. dalii* has not been found or described, except by Navás (1935) who gives a very poor figure of a larva which he believes to be that of *S. dalii*. Withycombe (1923) states that he had seen larvae of *S. terminalis* and *S. fuscata*, but he points out no differences of any kind between the two species. Larvae of the relatively common *S. fuscata* (Figs 10B, 11) have been found by several workers and described in detail by Withycombe (1923) and Killington (1936).

The life cycle of *S. fuscata* takes one year or less and larvae can be found from May to October, occasionally November, both on and inside sponges inhabiting lakes, canals and streams. The larvae leave the water in autumn and overwinter as prepupae in the pupal cocoon on a tree-trunk or wall. Length of larva is about 0·5 mm in the first instar and about 5 mm in the third and final instar.

Little is known about the habitat of the other two species, but Withycombe (1923) found larvae of *S. terminalis* in sponges inhabiting slow streams, ponds and canals. Captures of adults of *S. dalii* indicate that larvae of this species probably occur in the same places as *S. fuscata*.

LIFE CYCLES AND ECOLOGY

SIALIS

The family Sialidae has a wide distribution, occurring in Europe, America, Asia Minor, Siberia and Japan. Six species, all in the genus *Sialis*, occur in Europe but only *Sialis lutaria* and *S. fuliginosa* have been found in Britain.* Many workers have studied the general biology of *S. lutaria* but comparatively little is known about *S. fuliginosa*.

Sialis lutaria

The large number of studies on the internal anatomy and physiology include work on the development of the embryo (Du Bois 1936), the circulatory system of adults and larvae (Selman 1965), the morphology and behaviour of the blood cells (Selman 1960a, b, 1962), the blood composition, ionic regulation and water balance in the larva (Beadle & Shaw 1950, Shaw 1955a, b, Sutcliffe 1962, 1963), the excretion of ammonia by the larva (Staddon 1955), and the physiological changes during pupation (Geigy 1948, Geigy & Ochsé 1940a, b, Ochsé 1944, Geigy & Rahm 1951, Rahm 1952, Selman 1960a).

Sialis lutaria is a fairly widespread species and the larvae live in ponds, lakes and sluggish parts of streams and rivers where there is an abundance of silt. They are often numerous in the benthos, especially in humic acid lakes and ponds (Berg & Petersen 1956, Matthey 1971, McLachlan & McLachlan 1975). The name *lutaria* is derived from the habitat of the larva (Latin *lutum* = mud). Detailed studies of the life cycle have been made in Denmark (Berg 1938, Kaiser 1950, 1961), France (Giani & Laville 1973), Germany (Seitz 1940) and Switzerland (Du Bois & Geigy 1935, Geigy & Du Bois 1935, Geigy 1937, Geigy & Grobe 1958). There are only brief accounts of the life cycle in Britain (McLachlan 1868, Killington 1929, Syms 1935). The life cycle usually takes about two years from egg to adult but a longer period of three years has been recorded in lakes at high altitudes in Switzerland (Geigy & Grobe 1958) and the French Pyrenees (Giani & Laville 1973).

---

*A third species, *S. nigripes*, has now been found in Britain, (see footnote on p. 14).

The flight period in Britain is from late April to the end of June but adults have occasionally been taken in July in the south of England (Killington 1926). Their flight is very slow and laboured, and when disturbed, they usually run rather than fly. Females are much larger than males, e.g. mean and range for body length in a hundred adults of each sex from various localities in the Lake District were 10 mm (7-13 mm) for males and 13 mm (11-16 mm) for females. Although the adults have biting mouthparts, there are no definite records that they feed. Du Bois & Geigy (1935) found pollen on the head and thorax, and therefore concluded that the adults visit flowers to feed. Kaiser (1961) observed females apparently looking for food on flowers of cow parsley (*Anthriscus silvestris* (L.) Hoffm.). Similar behaviour was observed in two North American species, *S. rotunda* Banks and *S. californica* Banks, by Azam & Anderson (1969) but these authors found that the alimentary canal was not developed and therefore feeding was doubtful. Adults of another North American species, *S. cornuta* Ross, fed on sugar solution in the laboratory and were kept alive for up to one week (Pritchard & Leischner 1973). The life of the adults is short, usually one week and rarely longer than two weeks. Dead or dying adults frequently fall into the water and are eaten by fish, some adults are eaten by birds and spiders, and dead males are sometimes found in a mummified condition clinging to plants (Du Bois & Geigy 1935).

Adults are most active on warm sunny days, especially in the morning. Large females are sometimes seen in flight pursued by one or more smaller males. Mating always occurs on the ground, usually in the morning, whereas oviposition can occur throughout the day from mid-morning (about 11.00 h) to evening. The female attracts the male by producing a scent for which the receptor is on the male antennae (Geigy & Du Bois 1935), and often a female is surrounded by several males. Courtship usually starts with one male confronting the female and both vibrating their antennae for a few seconds. The female then walks a short distance with the male following. When the female stops or slows down, the male vibrates his antennae and mouthparts over the hind legs, wing tips and end of the abdomen of the female. The antennae and notched labrum of the male are thought to be receptor organs (Geigy & Du Bois 1935). Rupprecht (1975) has found that males and unmated females are able to find each other and communicate by generating vibration-signals. These are produced by rhythmic movements of the abdomen in both sexes and are transmitted through the legs and the ground to receptors on the legs of the opposite sex. If the female is ready for copulation, the male forces up the end of her abdomen with his head and moves underneath. He then turns his abdomen under one pair of wings and curls it

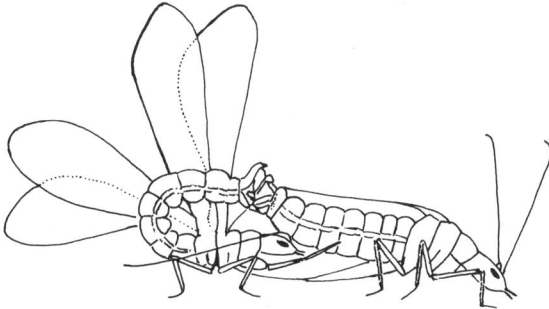

Fig. 13. Mating in *Sialis lutaria*; male is on the left, female on the right (cf. Plate 1).

over dorsally so that the genitalia meet (Fig. 13 and Plate 1). The pair remain in copulation for several minutes and a large white spermatophore is transferred from male to female. After one or two twists of the abdomen, the male disengages, leaving the spermatophore held ventrally by the genitalia of the female. After walking a short distance, the female spreads her legs wide apart, curls her abdomen under her thorax and chews on the spermatophore. The female copulates only once.

Fig. 14. Eggs of *Sialis lutaria*, length with micropyle about 0·8 mm.

Soon after feeding on the spermatophore, the female searches for an oviposition site. The eggs are usually laid on the stems and leaves of plants overhanging the water but are occasionally found on stones, tree trunks and bridges. Each egg is brown, cylindrical, with a short white cylindrical micropylar knob on the upper end (Fig. 14). The eggs are arranged in a single layer and at a slight angle to the surface on which they are attached (see cover and Plate 1). The female lays the eggs in rows and one female can lay between 300 and 900 eggs in a single mass

(Du Bois & Geigy 1935, Seitz 1940, Kaiser 1961). Sometimes a second, smaller mass of eggs is laid after a rest period of about six days. Lestage (1920a) gives a detailed description of oviposition behaviour in *S. lutaria*.

The eggs are frequently parasitized by a minute hymenopteron which was named *Oophthora semblidis* by its discoverer Aurivillius (1898) and later called *Pentarthron semblidis* by Lestage (1919) and *Trichogramma evanescens* Westw. by Kryger (1919), Lestage (1920a) and Du Bois & Geigy (1935). From his detailed study of the parasite, Salt (1937) concluded that the correct name was *Trichogramma semblidis* (Aurivillius). Salt collected nearly a quarter of a million eggs of *S. lutaria* and found that about 0·6% of them were attacked by the parasite. Much higher percentages have been reported for this species parasitizing eggs of North American species, with values of 14% for *S. rotunda*, 37-65% for *S. californica* (Azam & Anderson 1969), and 78% for *S. cornuta* (Pritchard & Leischner 1973). It is easy to recognize eggs that were parasitized because the emerging parasite bites a small round hole, whereas the *Sialis* larva makes a longitudinal slit on hatching.

The development of the embryo has been studied in detail by Du Bois (1936) and Seitz (1940). Development time varies from about seven days at 24°C to about twenty days at 15°C and few eggs develop at temperatures below about 10°C (Seitz 1940). The eggs usually hatch at night and nearly all larvae from one egg mass emerge simultaneously. Most eggs hatch in May and June but a few may still be hatching in early July. An egg burster on the cuticle of the embryo is used to cut a slit through the egg shell. The first instar larvae fall into the water or on to the ground and then crawl into the water. They are between 0·7 mm and 2·0 mm long and are active swimmers with long hairs on the legs, abdomen and gills (Fig. 15). Their appearance is very different from the other larval instars (Fig. 9), and both Lestage (1919) and Bertrand (1949) have compared the morphology of the two larval forms. As the first instars are essentially planktonic, they disperse rapidly from the comparatively limited area where newly-hatched larvae enter the water. They are positively phototactic whereas all other larval instars are negatively phototactic. The first instar moults after three to five days and the negatively photo-tactic second instar descends to the bottom and burrows in the mud.

There are ten larval instars, and this appears to be the usual number for most species of *Sialis*, but a North American species, *S. cornuta*, pupates at any instar between seven and ten (Pritchard & Leischner 1973). Larvae of any instar of *S. lutaria* can be induced to pupate by experimental and surgical techniques, but the percentage success increases with the instar number (Selman 1960a). The size of each larval instar varies

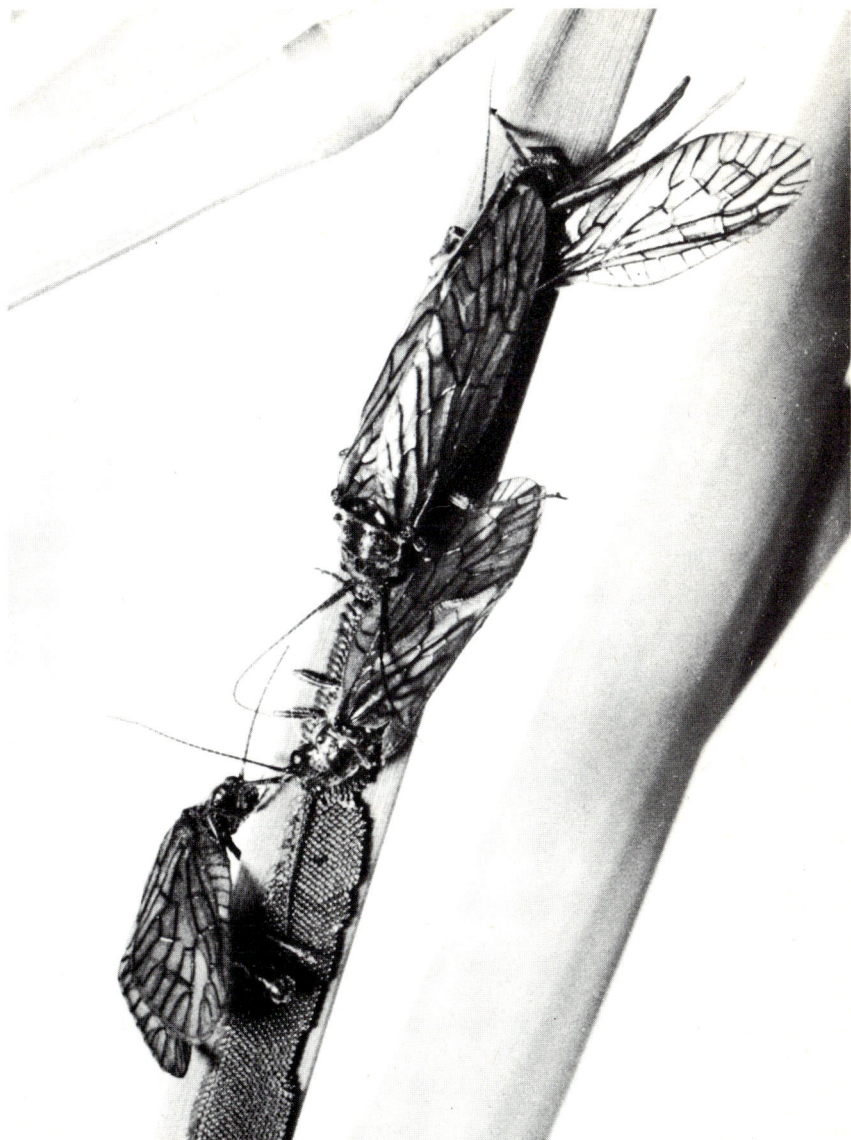

Plate 1. Eggs and mating in *Sialis lutaria*. (Photo: J. M. Elliott)

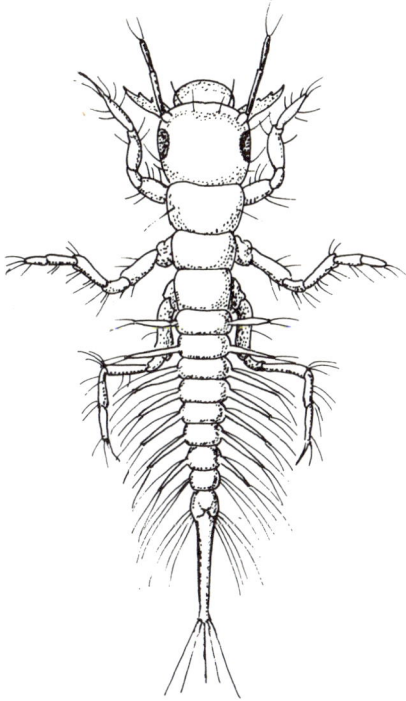

Fig. 15. First instar larva of *Sialis lutaria*, overall length 0·7–2·0 mm.

slightly from one locality to another and there is a marked sexual dimorphism in size in the later instars. For example, ranges for body length including the terminal filament are usually 14-22 mm in male larvae and 18-26 mm in female larvae. Although body length can be used to separate the ten instars, a more precise character is the width of the head capsule (Giani & Laville 1973).

Most larvae are in the eighth or ninth instar by the end of their first year (May or June) and in the tenth instar in autumn (September or October). They usually pass their second winter in the tenth instar and finally leave the water for pupation in late March, April, May or the beginning of June. Therefore the larval stage usually lasts about two years, but a longer period of three years has been found in lakes at high altitudes, with the larvae in only the fifth or sixth instar at the end of their first year (July or August) and in the eighth or ninth instar at the end of their second year (Geigy & Grobe 1958, Giani & Laville 1973).

The larvae are carnivores and their predominant food organisms are chironomid larvae and oligochaetes in the larger larvae (Griffiths 1973, Giani & Laville 1973), benthic crustaceans in the smaller larvae (Giani & Laville 1973) and micro-organisms and detritus in the first instar larvae (Du Bois & Geigy 1935). Older larvae (instars seven to ten) sometimes eat younger larvae. Food organisms are usually swallowed whole but larger prey are manipulated into the foregut with the forelegs and mandibles, and then bitten into two pieces. The prey is retained in the foregut by proventricular teeth whilst enzymes from the midgut initiate breakdown of the food into fine particles which pass into the midgut. Digestion chiefly occurs in the foregut and undigested material passes rapidly through the midgut to form faecal pellets. Pritchard & Leischner (1973) found that the mean minimum time between ingestion of prey and evacuation of the first faecal pellet in S. cornuta was 106 h at 5 °C, 29 h at 12·5 °C and 22 h at 17·5 °C. Larvae are eaten by other carnivorous invertebrates such as larvae of Odonata (Griffiths 1973), and also by fish such as brown trout, Salmo trutta L. (e.g. Macan 1966, Giani & Laville 1973).

The larvae occur in the littoral, sub-littoral and sometimes the profundal zones of lakes and have been recorded at depths down to 7 m in Lake Gribsø (Berg & Petersen 1956), 12 m in Windermere (Humphries 1936, Macan & Worthington 1972) and Lake Esrom (Berg 1938), about 13 m in the Plöner lakes (Lundbeck 1926), 15 m in lakes Fure and Hald (Wesenberg-Lund 1917) and Lake Port-Bielh (Giani & Laville 1973), and 20 m in Sempachersee (Du Bois & Geigy 1935). All instars do not have the same distribution pattern on the bottom. As the larvae grow, their maximum numbers are found at progressively greater depths. The smaller larvae in their first year are therefore most abundant in the littoral, whereas the larger larvae in their second year are most abundant in the sub-littoral or profundal zones, except when they migrate inshore before pupation in spring (Du Bois & Geigy 1935, Berg & Petersen 1956, Giani & Laville 1973).

Tenth instar larvae leave the water in late March, April, May or early June, the exact time varying from one locality to another. Larvae usually move on to land in the evening and early hours of the night, and often move some distance from the water's edge before they find a suitable site for pupation. Miall (1895) describes how a larva climbed a concrete wall, passed through a thicket of Cotoneaster and finally pupated in a flower bed that was six yards from the water. The tracheal gills of the larva adhere to the body instead of projecting outwards and their contents are retracted to form a small lump at the base of each gill.

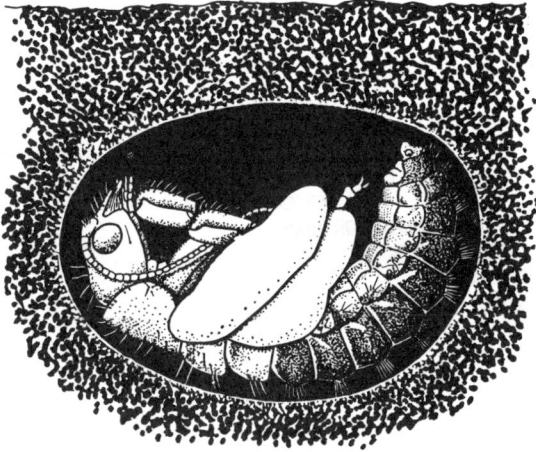

Fig. 16. Pupal chamber and pupa of *Sialis lutaria*.

Pupation usually occurs in fairly damp loose soil or in vegetable debris in the middle of clumps of sedge (Killington 1926). Each larva digs a pupal chamber about one centimetre below the soil surface and within a day it has moulted into a pupa. No cocoon is spun in the pupal chamber and the pupa has the legs and rudimentary wings free. The body is curved inwards and the abdomen has spines which keep the pupa free from direct contact with the walls of the chamber. In Fig. 16, the pupa is shown lying horizontally in a closed chamber just below the soil surface, but Jacques (1974) notes that the pupae lie vertically, head uppermost, in open chambers. In his detailed study of pupation, Seitz (1940) shows that there are actually three types of pupal chamber, an open-mouthed chamber in which the larva lies vertically, a J-shaped chamber which is open at one end with the larva lying vertically in the closed limb, and a closed chamber as in Fig. 16. He found between 30 and 40% of the pupae in closed chambers which are made only in firm soil. The pupal stage requires a total of 72·5 degree-days above a base temperature of 6·6°C (Kaiser 1961), e.g. it lasts 30 days at 9·0°C, 20 days at 10·2°C, 10 days at 13·9°C and 5 days at 21·1°C. Emergence of the adult usually occurs soon after dawn in late April, May, June and occasionally July. If the pupal chamber is open, the final moult occurs in the chamber, where the cast pupal skin is left, but if the chamber is closed, the pupa digs its way out and the final moult occurs on the soil surface.

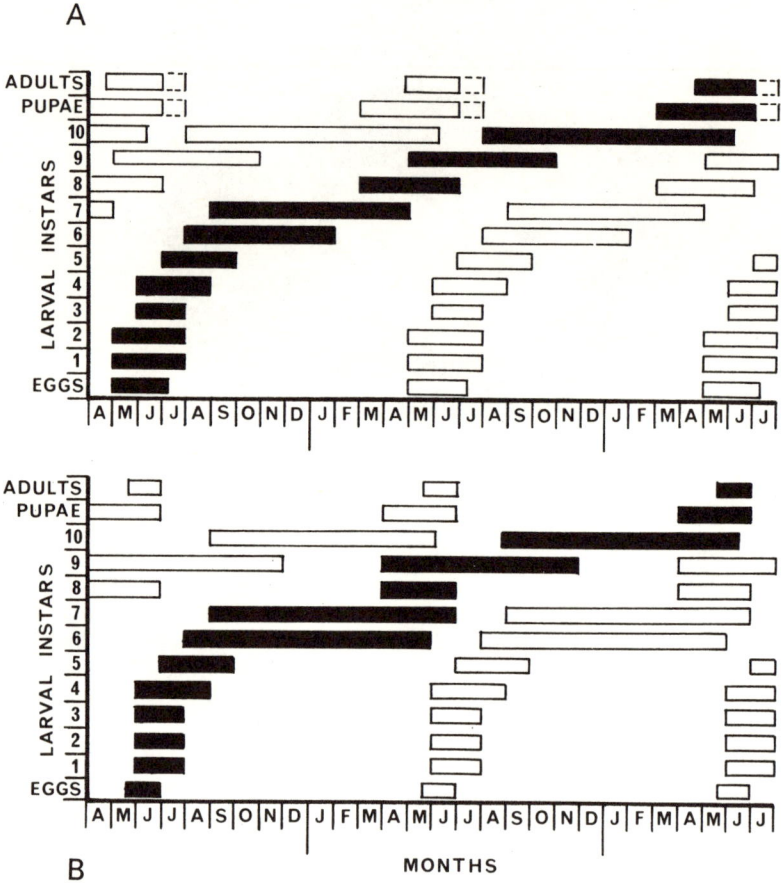

Fig. 17. Life cycle of: A, *Sialis lutaria*; B, *Sialis fuliginosa*. Each horizontal bar indicates the months in which the various stages of the life cycle occur.

The life cycle is thus complete and usually takes two years. A summary of the two-year cycle is shown in Fig. 17A. As the actual months in which each instar occurs will vary from one locality to another, this figure can only serve as a guide.

The common name of the adult *Sialis* is the alder or orl fly, and an imitation of the adult is used by fly-fishermen during the flight period from late April to June. Adults are often found in the stomachs of trout and are usually moribund adults that have fallen into the water after mating or oviposition. As they are poor fliers, some adults may become fatigued at some distance from the land, especially in a strong wind, and may then fall or be blown on to the water surface.

Sialis fuliginosa

Nothing is known about the internal anatomy and physiology of *S. fuliginosa* and very little is known about its ecology. The name *fuliginosa* (from Latin *fuligo* = soot or black paint) probably refers to the wings, which are blacker than those of *S. lutaria*. The larvae are limited to moderately fast streams and the upper reaches of rivers. As larvae of *S. lutaria* are sometimes found in rivers, there may be some places where the two species occur together, but nothing is known about any possible competition between them.

There are some detailed observations on the life cycle in Denmark (Kaiser 1950, 1961), and the following account is based on this work and some additional observations made in two small stony streams, the Walla Brook on Dartmoor in southwest England and the Wilfin Beck in the English Lake District. These streams are described in detail by Elliott (1967a, 1973). The life cycle takes two years.

The flight period starts slightly later than that of *S. lutaria* and usually lasts about six weeks in May, June and occasionally July. Adults are slightly larger than those of *S. lutaria*, e.g. mean and range for body length in adults from the Walla Brook and Wilfin Beck were 12 mm (8-15 mm) for males and 16 mm (12-20 mm) for females. Mating has been described by Killington (1932) and is very similar to that of *S. lutaria* except that the male holds the abdomen of the female with one or both forelegs. Rupprecht (1975) has found that males and unmated females generate vibration-signals similar to those produced by *S. lutaria*, except that the vibrations are generated by tapping the abdomen and wings on the ground. Kaiser (1961) notes that the eggs are never laid on vegetation along the banks of the stream and are usually found about 3-5 m above the surface of the water on leaves of deciduous trees. The eggs are light brown but are otherwise similar to those of *S. lutaria*. They are usually laid in a circular mass containing between 300 and 1100 eggs. Each egg mass is laid by one female and the smaller egg masses (300-400 eggs) are probably from females that have oviposited for a second time. The development time in the egg stage is about the same as that of *S. lutaria* (Kaiser 1961) and the first instar larvae are similar to those of *S. lutaria*.

There are ten larval instars and the fully-grown larvae are larger than those of *S. lutaria*, e.g. body length including the terminal filament is usually between 18 and 26 mm in male larvae and 23 and 32 mm in female larvae.   Most larvae are in the sixth, seventh or eighth instar by the end of their first year (May or June) and in the tenth instar in autumn (September or October).   They finally leave the water for pupation in late April, May or June.   Hildrew & Townsend (1976) found that the larvae took prey items roughly in proportion to their numerical abundance in the benthos, the chief food organisms being chironomid larvae and plecopteran nymphs.   The larvae are more active than those of *S. lutaria* and often hunt their prey.   They are themselves eaten by brown trout (Elliott 1967b).

Pupation and the final emergence of the adult are the same as in *S. lutaria*, except that the pupal stage requires a slightly higher total of 84·5 degree-days above a base temperature of 7·2°C (Kaiser 1961), e.g. it lasts 30 days at 10°C, 20 days at 11·4°C, 10 days at 15·7°C and 5 days at 24·1°C.   A summary of the two-year life cycle in the Walla Brook and Wilfin Beck is shown in Fig. 17B.

Osmylus

The family Osmylidae has a wide distribution, occurring in Europe, Asia, Africa, New Zealand, Australia and South America.   Two species, both in the genus *Osmylus*, occur in Europe but one species, *O. elegantissimus* K., is restricted to the Caucasian region whilst the other, *O. fulvicephalus*, occurs throughout most of Europe, including Britain (Illies 1967).

Osmylus fulvicephalus

This is a local species but usually forms colonies so that it is fairly abundant where it is found.   Adults are often found under bridges and near woodland streams with a dense cover of overhanging vegetation on their banks.   Larvae live in damp moss at the edge of the stream or on overhanging banks.   Early observations on the life cycle were made by Stein (1838), Brauer (1850, 1851), Hagen (1852) and McLachlan (1868), but more detailed accounts are provided by Withycombe (1923), David (1936), Killington (1936) and Ward (1965).

The adults are the largest of the British Neuroptera (Fig. 3C).   Their flight period is usually from May to July, but they have occasionally been taken in April and August.   They are feeble flyers and are most active at dusk when they feed on small insects including aphids (Withycombe 1923) and adult Lepidoptera which are probably caught when resting, injured or even dead (Killington 1936). The length of life of a male is only a few weeks but females can live for as long as two to three months (Withycombe 1923).

Courtship is unusual because the male is the more passive sex and attracts the female. He has a pair of whitish, finger-shaped, eversible scent-glands between the eighth and ninth abdominal segments and these can be extruded to a length of about 4 mm. Withycombe (1923) first described the courtship and mating in detail. When the male is ready for mating, he hangs down from a support, lifts his wings and everts his scent-glands. Females in the immediate vicinity respond almost at once by waving their antennae vigorously, then flying or walking towards the male and stroking the scent-glands with their antennae and palps. The male withdraws the scent-glands and turns to meet the female. After male and female have stroked their antennae for a few minutes, the male carefully bites on one of the female's fore-coxae, each of which carries a teat-like process. Both insects bend their abdomens to meet, the apex of the female's abdomen curling over and then under that of the male so that the elongate ventral valves of the female grasp the genitalia of the male. The male now releases his hold on the female and the ventral valves of the female lever on the male genitalia for between ten and sixty minutes until a large white spermatophore is produced by the male. After the separation of the two insects, the spermatophore is seen projecting from the apex of the female's abdomen. The spermatophore is about 4 mm long, white in colour but yellowish at the centre, and consists of four rounded lobes and a short attaching stem. Soon after the female has separated from the male, she bends her head under her body and eats the nearest lobe of the spermatophore. If she continues to eat the spermatophore, the male intervenes and caresses her. When the female finally leaves the male, the remaining part of the spermatophore is either eaten or becomes dislodged. A female may mate two or three times but once is sufficient for all the eggs laid to be fertile.

Eggs are laid two to three days after mating in late May, June and early July. Each female lays about thirty eggs in small batches on leaves, twigs and especially the leaves of moss near the water (Lestage 1920b, David 1936, Ward 1965). The eggs are cylindrical, slightly flattened, with a short micropylar knob, a length of between 1·6 and 1·8 mm and a breadth of between 0·5 and 0·6 mm. They are glossy-white when laid but darken to creamy-yellow or brown within a few days. Eggs deposited on moss are laid singly or in pairs on the underside of leaves with the micropyle towards the tip of the leaf. They are also laid in regular, straight or slightly curved rows on larger objects, each row with two to twenty eggs touching one another. The eggs hatch within four to twenty-two days, depending upon the temperature (Withycombe 1923, David 1936, Ward 1965), and the larva escapes through a slit made by a very long, saw-like egg-breaker. Within one to three days after emergence, the larvae leave the locality of the eggs and burrow into the moss.

There are only three larval instars. The first instar is slightly different from the other two instars; the head is broader than the thorax, the jaws, setae and legs are relatively longer than those of later instars, and the fore-gut contains one or two bubbles of air which may prevent the larva from sinking under water (Lestage 1920b, Withycombe 1923). Second and third instars are very similar (Fig. 10A). Ward (1965) measured large numbers of captive larvae at intervals of one to three days and obtained the following mean values and ranges: overall length (inclusive of jaws), first instar 4·95 mm (3·53-5·93 mm), second instar 8·91 mm (7·14-10·39 mm), third instar 15·70 mm (13·50-16·20 mm); width of head capsule, first instar 0·54 mm (0·50-0·59 mm), second instar 0·86 mm (0·80-0·88 mm), third instar 1·54 (1·53-1·56 mm).

There is some disagreement over whether the larvae are truly aquatic, semi-aquatic or terrestrial. Several workers have stated that the larvae are definitely aquatic (e.g. Brauer 1850, 1851, Girard 1879, Withycombe 1923), but Hagen (1852) and Brocher (1913) considered that they were terrestrial. Other authors have found that the larvae live in damp moss and debris near the water but may enter the water in search of prey (McLachlan 1868, David 1936, Killington 1936). Ward (1965) found that when larvae were dropped into water, they rapidly crawled out, and that when larvae inside moss were immersed in water, they crawled deep into the moss and died within eight to twenty-eight days. He therefore concluded that although the larvae live near the water's edge, they never enter the water of their own accord and cannot be regarded as aquatic or even semi-aquatic.

The larvae are carnivores and first instars feed chiefly on small arthropods, such as mites and Collembola, whereas second and third instars feed chiefly on dipterous larvae, such as Chironomidae and Tipulidae. If a movement is detected in the moss or mud on which the larva is walking, the larva stabs downwards with its long, slender jaws. When the tips of the jaws strike a prey, a salivary secretion is injected into the prey which is quickly paralyzed. The contents of the prey are then slowly removed by the sucking jaws. The paralyzing effect of the secretion is far more rapid in chironomid larvae than in surface-dwelling arthropods such as aphids and mites, and a large chironomid larva is paralyzed after only ten seconds (Withycombe 1923).

The growth rate of the larvae depends upon the food supply and temperature. Ward (1965) found that the duration of each instar in captivity was between fourteen and twenty-four days for first instars, twenty-eight and thirty-nine days for second instars, and 248 days for a third instar larva. All larvae are in the second or third instar by the end of the summer. In late September or October, the larvae cease to feed, and burrow deep into the roots of the moss where they overwinter in a state of hibernation known as diapause. Larvae in this condition can survive total immersion in water and this may happen frequently during winter spates (David 1936). Hibernation ends in March and the larvae start to feed again.

In late April or early May, a silken cocoon is spun amongst the moss, some moss being incorporated into the cocoon. The cocoon is a thin, closely woven, yellowish-white structure, irregularly oval in shape, about 10 mm long and 8 mm broad. The larva remains in a prepupal stage in the cocoon for seven to eighteen days and then pupates. Just prior to pupation, the long jaws of the larva break off near their base. The pupal appendages are free but immovable until just before emergence. There is a pair of well-developed pupal mandibles and the head and abdomen are bent ventrally. The pupal stage lasts about ten to fourteen days and then the pupa cuts an irregular slit in the cocoon with its mandibles. After the pupa has crawled to some firm support, the final moult occurs, usually in the morning or evening, and the adult emerges.

A summary of the one-year life cycle is shown in Fig. 18A. As the life-cycle will vary from one locality to another, this figure can only serve as a guide.

SISYRA

The family Sisyridae has a wide distribution, occurring in Europe, Asia, Africa, North America, South America, the West Indies, the Philippine Islands and Australia. Taxonomic works on the whole family include monographs by Krüger (1923), Navás (1935) (reviewed by Lestage 1935), and Parfin & Gurney (1956). The family appears to be closely related to the Osmylidae in spite of the very different appearance of the imagines. Withycombe (1925) considered that the Sisyridae were derived from osmyloid ancestors, the larvae of which took to deep water and then to preying on freshwater sponges. He suggested that the larvae developed longer and slimmer jaws, and abdominal tracheal gills, that their labial palpi were lost and that the two claws on each tarsus fused laterally into a single claw.

Twenty-three species have now been recorded in the genus *Sisyra*. Only five species occur in Europe and three of these species have been found in Britain. Several workers have studied the life cycle and general biology of *S. fuscata*, but comparatively little is known about *S. terminalis* and *S. dalii*. The following account is based chiefly on those of Withycombe (1923, 1925) and Killington (1936.)

Sisyra fuscata.

This species is widely distributed and is the only sisyrid known to occur in both the Nearctic and Palaearctic regions. Larvae are found both on and inside sponges inhabiting lakes, canals, rivers and streams, and the adults are usually found on the vegetation near the water. The life cycle takes one year or less, and in some localities there may be two broods in one year.

The flight period is from May to September but adults are most abundant in May and June. Eggs are laid in May and June and most of the larvae from these eggs grow slowly through the summer, leave the water in autumn, overwinter in the prepupal stage within the cocoon, and pupate in April, May and June. The adults finally emerge in May and June to complete a one-year cycle. A small number of larvae from the eggs laid in May and June grow rapidly to reach the pupal stage in a few weeks and the adult stage in August and September to complete the life cycle in four to five months. These adults lay eggs in August and September, the larvae grow rapidly, overwinter in the prepupal stage and pupate in April, May and June. This complex life cycle is summarized in Fig. 18B, but it must be remembered that the life cycle will vary from one locality to another, and there may be no rapidly-growing second brood in some localities.

The adults are most active in the evening and sometimes form small swarms over the water. Little is known about their food but Tjeder (1944) observed an adult male feeding on eggs of *Sialis lutaria*, and other sisyrids feed upon plant products such as nectar (Brown 1952). The adults are eaten by birds, spiders and predaceous insects, and Berg (1948) found adults infested with larvae of the mite family Trombidiidae.

Mating occurs at dusk and several hours to two weeks after emergence. There is no elaborate courtship and the male copulates laterally with the female. After about three to five minutes, a small white spermatophore is left attached to the tip of the abdomen of the female and is immediately eaten. Oviposition takes place in the evening, often only a few hours after pairing, and the eggs are laid in small clusters of from one to twelve eggs. The blade-shaped ninth sternite (Fig. 7E) contains the opening of the cement gland and is also used for locating crevices in which

A

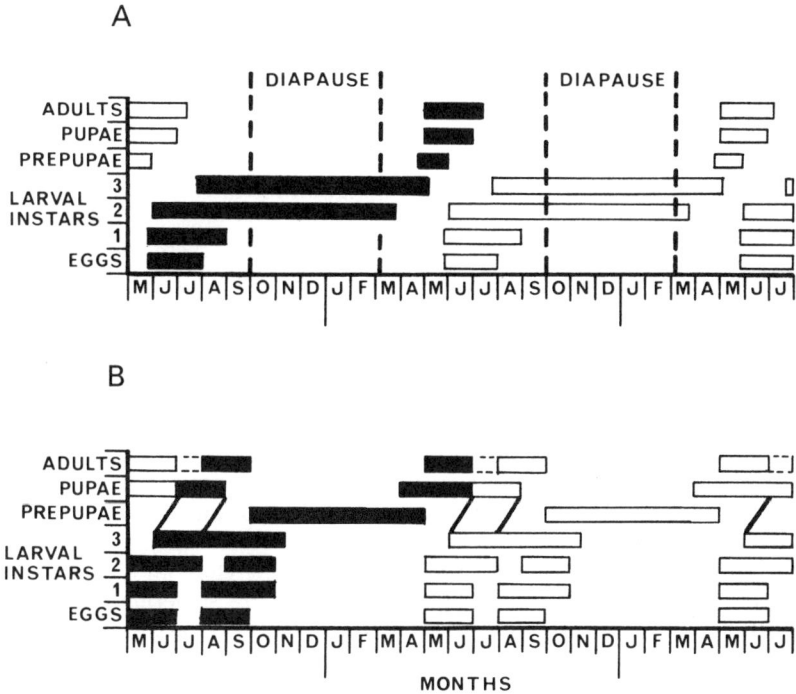

B

Fig. 18. Life cycle of: A, *Osmylus fulvicephalus*; B, *Sisyra fuscata*. Each horizontal bar indicates the months in which the various stages of the life cycle occur.

to deposit eggs on leaves, wooden piles and other objects overhanging the water. Each batch of eggs is covered by a web of white silk. The female spins a few parallel strands, then shifts her position slightly and spins a few more strands, crossing the first strands, and so on until three or four layers of silk are laid down. The eggs are pale yellow, elongate oval, with a small flattened micropylar knob and a length of about 0·35 mm.

Hatching usually occurs within eight to fourteen days, the incubation period being partially dependent upon the temperature. The larva uses its egg-breaker to saw a slit in the egg shell, works its way out of the shell, pushes through the layer of silk and falls on to the surface of the water. It then penetrates the surface film by bending the tip of the abdomen back over the dorsal side of the head, so that the anterior part of the body is forced under the water as the body is suddenly straightened. The submerged larva floats in the water, occasionally swims, and drifts about until it comes into contact with a sponge. It at once attaches to the sponge and starts to feed by inserting its jaws and sucking the fluids of the sponge. The usual hosts are *Spongilla lacustris* (L.) and *Ephydatia fluviatilis* (L.), and sisyrid larvae have been found only on members of the family Spongillidae. Larvae have also been found on filamentous algae and bryozoans (Wesenberg-Lund 1943).

There are only three larval instars. The first instar larva is slightly different from those of the other two instars; the jaws are stouter and shorter, the abdomen is narrower than the thorax, there are very long setae on the ninth abdominal segment and abdominal gills are apparently absent. Second and third instars are very similar (Fig. 10B), and have seven pairs of segmentally-arranged jointed gills on the ventral surface of the abdomen (Fig. 11). The body length, excluding the jaws, is about 0·5 mm in the first instar, about 1·5 mm in the second instar and between 3 and 6·5 mm in the third instar.

The complex life cycle has been described above. Fully-fed larvae in the third instar swim away from the sponge to the shore. Whilst swimming, they are rather conspicuous and must be an easy prey for fish. On leaving the water, they frequently travel a considerable distance before selecting a site for spinning a cocoon, and are therefore an obvious prey for predatory arthropods and birds. Cocoons are sometimes found in great numbers on the walls of bridges and in crevices of bark or under loose bark on tree trunks. The cocoon has a double structure; an outer, coarse, yellowish silk network and an inner, finer, white cocoon. When placed in a crevice, the outer envelope forms a canopy over the inner, the whole structure being about 4 mm by 3·5 mm. The larvae that leave the water in autumn spend the winter as resting larvae (prepupae) in the pupal cocoon and finally pupate in early spring. Larvae leaving the water in mid-summer pupate immediately. Large numbers of prepupae and a few pupae are sometimes killed by a white mould (Killington 1936), and a parasitic hymenopteron, *Eupteromalus* sp., which lays an egg beside the resting larva (Killington 1933).

The pupa is short and stout with large mandibles and free appendages. After a short pupal period of about two weeks, an irregular hole is cut through the anterior end of the cocoon by the pupal mandibles. The pupa does not always leave the cocoon before the emergence of the adult. Most adults emerge in the evening, after sunset.

Sisyra terminalis and *S. dalii*

Very little is known about these species. Withycombe (1923) states that he had seen all stages of *S. terminalis* and that they are hardly distinguishable from those of *S. fuscata*. He also records that the larvae occur in sponges in streams, canals and ponds. As adults of this species have been taken in May, June and July, but not in late summer, there does not appear to be a second brood in late summer, as seen in *S. fuscata*.

Even less is known about *S. dalii*, except that it appears to occur in the same places as *S. fuscata*. As adults have been taken in June, July and August, the various stages of the life cycle probably occur at slightly later times than those of *S. fuscata*, and there is probably no second brood in late summer.

ACKNOWLEDGMENTS

I wish to express my sincere thanks to the following for their help in the preparation of this booklet: Mrs Joan Worthington for her excellent illustrations; Mrs P. A. Tullett for all her assistance in many aspects of this work; J. E. M. Horne for editing the manuscript and supervising its conversion to the final booklet; E. W. Kaiser of Klokkedal, Denmark for many helpful suggestions and for sharing his extensive knowledge of the genus *Sialis*: Dr P. C. Barnard of the British Museum (Natural History) for helpful advice and for the loan of specimens; Dr P. D. Armitage, Dr A. G. Hildrew, R. A. Jenkins, Dr M. A. Learner, Mrs M. J. Morgan, Dr R. S. Wilson for supplying specimens.

REFERENCES

Aurivillius, C. (1898). En ny Svensk äggparasit. *Ent. Tidskr.* **18**, 249-56.

Azam, K. M. & Anderson, N. H. (1969). Life history and habits of *Sialis rotunda* and *S. californica* in Western Oregon. *Ann. ent. Soc. Am.*, **62**, 549-558.

Beadle, L. C. & Shaw, J. (1950). The retention of salt and the regulation of the non-protein nitrogen fraction in the blood of the aquatic larva, *Sialis lutaria. J. exp. Biol.*, **27**, 96-109.

Berg, K. (1938). Studies on the bottom animals of Esrom Lake. *K. danske Vidensk. Selsk. Skr. (naturv. math.)*, ser. 9, **8**, 1-255+Pl. 1-17.

Berg, K. (1948). Biological studies on the River Susaa. *Folia limnol. scand.*, **4**, 1-318.

Berg, K. & Petersen, I. B. (1956). Studies on the humic acid Lake Gribsø. *Folia limnol. scand.*, **8**, 1-273.

Bertrand, H. (1949). Notes morphologiques sur les larves des *Sialis* L. (Mégaloptères, Sialidae). *Feuille Nat.*, **4**, 5-10.

Brauer, F. (1850). Ueber die Verwandlung des *Osmylus maculatus. Ber. Mitt. Freunden Naturw. Wien*, **7**, 153.

Brauer, F. (1851). Verwandlungsgeschichte des *Osmylus maculatus. Arch. Naturgesch.*, **17**, 255-258.

Brocher, F. (1913). *L'aquarium de chambre.* Lausanne, 317 pp.

Brown, H. P. (1952). The life history of *Climacia areolaris* (Hagen), a neuropterous 'parasite' of freshwater sponges. *Am. Midl. Nat.*, **47**, 130-160.

David, K. (1936). Beiträge zur Anatomie und Lebensgeschichte von *Osmylus chrysops* L. *Z. Morph. Ökol. Tiere*, **31**, 151-206.

Du Bois, A. M. (1936). Recherches expérimentales sur la détermination de l'embryon dans l'oeuf de *Sialis lutaria. Revue suisse Zool.*, **43**, 519-523.

Du Bois, A. M. & Geigy R. (1935). Beiträge zur Ökologie, Fortpflanzung-biologie und Metamorphose von *Sialis lutaria* L. *Revue suisse Zool.*, **42**, 169-248.

Elliott, J. M. (1967a). Invertebrate drift in a Dartmoor stream. *Arch. Hydrobiol.*, **63**, 202-237.

Elliott, J. M. (1967b). The food of trout (*Salmo trutta*) in a Dartmoor stream. *J. appl. Ecol.*, **4**, 59-71.

Elliott, J. M. (1973). The life cycle and production of the leech *Erpobdella octoculata* (L.) (Hirudinea: Erpobdellidae) in a Lake District stream. *J. Anim. Ecol.*, **42**, 435-448.

Fraser, F. C. (1959). Mecoptera, Megaloptera and Neuroptera. *Handbk Ident. Br. Insects*, **1**, 12-13. 40 pp.

Geigy, R. (1937). Beobachtungen über die Metamorphose von *Sialis lutaria* L. *Mitt. schweiz. ent. Ges.*, **17**, 144-157.

Geigy, R. (1948). Etude expérimentale de la métamorphose de *Sialis lutaria* L. *Bull. biol. Fr. Belg.*, **33**, 62-67.

Geigy, R. & Du Bois, A. M. (1935). Sinnesphysiologische Beobachtungen über die Begattung von *Sialis lutaria* L. *Revue suisse Zool.*, **42**, 447-457.

Geigy, R. & Grobe, D. (1958). Die Ökologische Abhängigkeit des Metamorphose-Geschehens bei *Sialis lutaria* L. *Revue suisse Zool.*, **65**, 323-328.

Geigy, R. & Ochsé, W. (1940a). Schürungsversuche an larven von *Sialis lutaria* L. *Revue suisse Zool.*, **47**, 193-194.

Geigy, R. & Ochsé, W. (1940b). Versuche uber die inneren Faktoren der Verpuppung bei *Sialis lutaria* L. *Bull. biol. Fr. Belg.*, **47**, 225-241.

Geigy, R. & Rahm, U. H. (1951). Beiträge zur experimentellen Analyse der Metamorphose von *Sialis lutaria* L. *Bull. biol. Fr. Belg.*, **58**, 408-413.

Giani, N. & Laville, H. (1973). Cycle biologique et production de *Sialis lutaria* L. (Megaloptera) dans le lac de Port-Bielh (Pyrénées Centrales). *Annls Limnol.*, **9**, 45-61.

Girard, M. (1879). *Traité elementaire d'entomologie*, **2**, 443-456.

Griffiths, D. (1973). The food of animals in an acid moorland pond. *J. Anim. Ecol.*, **42**, 285-293.

Hagen, H. A. (1852). Die Entwicklung und der innere Bau von *Osmylus*. *Linnaea Ent., Berl.*, **7**, 368-418.

Hildrew, A. G. & Townsend, C. R. (1976). The distribution of two predators and their prey in an iron rich stream. *J. Anim. Ecol.*, **45**, 41-57.

Hoffmann, J. (1962). Faune des Neuroptéroides du Grand-Duché de Luxembourg. *Archs Inst. gr.-duc. Luxemb.*, **28**, 249-332.

Humphries, C. F. (1936). An investigation of the profundal and sublittoral fauna of Windermere. *J. Anim. Ecol.*, **5**, 29-52.

Illies, J. (Ed.) (1967). *Limnofauna Europaea.* Stuttgart. Fischer. 474 pp.

Jacques, D. (1974). *The development of modern stillwater fishing.* London. A. & C. Black, 237 pp.

Kaiser, E. W. (1950). *Sialis nigripes* Ed. Pict., new to Denmark, and the distribution of *S. lutaria* and *S. fuliginosa* Pict. in Denmark. *Flora Fauna, Silkeborg*, **56**, 17-36.

Kaiser, E. W. (1961). On the biology of *Sialis fuliginosa* Pict. and *S. nigripes* Ed. Pict. *Flora Fauna, Silkeborg*, **67**, 74-96.

Killington, F. J. (1926). Notes on Neuroptera taken in 1925. *Entomologist*, **59**, 110-112.

Killington, F. J. (1929). A synopsis of British Neuroptera. *Trans. ent. Soc. S. Engl.*, **5**, 1-36.

Killington, F. J. (1932). On the pairing of *Sialis fuliginosa* Pict. (Neuroptera: Megaloptera). *Entomologist*, **65**, 66-67.

Killington, F. J. (1933). The parasites of Neuroptera with special reference to those attacking British species. *Trans. ent. Soc. S. Engl.*, **8**, 84-91.

Killington, F. J. (1936). *A monograph of the British Neuroptera*, **1**. London. Ray Society. 269 pp.

Killington, F. J. (1937). *A monograph of the British Neuroptera*, **2**. London. Ray Society. 306 pp.

Kimmins, D. E. (1962). Keys to the British species of aquatic Megaloptera and Neuroptera with ecological notes. *Scient. Publs Freshwat. biol. Ass.* No. 8, 23 pp.

Krüger, L. (1923). Sisyridae. Beiträge zu einer Monographie der *Neuropteren*-Familie der Sisyriden. *Stettin. ent. Ztg,* **84,** 25-66.

Kryger, J. P. (1919). The European Trichogramminae. *Ent. Meddr,* **12,** 257-354.

Lestage, J. A. (1919). Notes biologiques sur *Sialis lutaria* L. (Megaloptera) *Annls Biol. lacustre,* **9,** 25-38.

Lestage, J. A. (1920a). Le méchanisme de la ponte chez *Sialis lutaria* L (Megaloptera) *Annls Biol. lacustre,* **10,** 221-223.

Lestage, J. A. (1920b). La ponte et la larvule de *l'Osmylus chrysops* L. (Planipenne). *Annls Biol. lacustre,* **10,** 226-230.

Lestage, J. A. (1935). Notes sur les sisyridés (Hémérobiiformes à larve aquatique). *Bull. Annls Soc. r. ent. Belg.,* **75,** 387-394.

Lundbeck, J. (1926). Die Bodentierwelt norddeutscher Seen. *Arch. Hydrobiol. (Suppl.),* **7,** 1-473.

Macan, T. T. (1966). The influence of predation on the fauna of a moorland fishpond. *Arch. Hydrobiol.,* **61,** 432-52.

Macan, T. T. & Worthington, E. B. (1972). *Life in lakes and rivers.* (Revised paperback edition). London. Fontana. 320 pp.

Matthey, W. (1971). Écologie des insectes aquatiques d'une tourbière du Haut-Jura. *Revue suisse Zool.,* **78,** 367-536.

McLachlan, A. J. & McLachlan, S. M. (1975). The physical environment and bottom fauna of a bog lake. *Arch. Hydrobiol.,* **76,** 198-217.

McLachlan, R. (1868). A monograph of the British Neuroptera— Planipennia. *Trans. Ent. Soc. Lond.,* 1868, 145-224.

Meinander, M. (1962). The Neuroptera and Mecoptera of Eastern Fennoscandia. *Fauna fenn.,* **13,** 1-96.

Miall, L. C. (1895). *The natural history of aquatic insects.* London. Macmillan, 395 pp.

Navás, L. (1935). Monografía de la Familia de los sisirídos (Insectos neuropteros). *Mems Acad. Cienc. exact. fís.-quím. nat. Zaragoza,* **4,** 1-86.

Ochsé, W. (1944). Experimentellen und histologische Beiträge zur inneren Metamorphose von *Sialis lutaria* L. *Revue suisse Zool.,* **51,** 1-82.

Parfin, S. I. & Gurney, A. B. (1956). The Spongilla-flies, with special reference to those of the Western Hemisphere (Sisyridae, Neuroptera) *Proc. U.S. natn. Mus.,* **105,** 421-529.

Pritchard, G. & Leischner, T. G. (1973). The life history and feeding habits of *Sialis cornuta* Ross in a series of abandoned beaver ponds (Insecta; Megaloptera). *Can. J. Zool.,* **51,** 121-131.

Rahm, V. H. (1952). Die innersekretorische Steuerung der postembryonalen Entwicklung von *Sialis lutaria* L. *Revue suisse Zool.,* **59,** 179-237.

Rupprecht, R. (1975). Die Kommunikation von *Sialis* (Megaloptera) durch Vibrationssignale. *J. Insect Physiol.,* **21,** 305-320.

Salt, G. (1937). The egg parasite of *Sialis lutaria*: a study of the influence of the host upon a dimorphic parasite. *Parasitology*, **29**, 539-553.

Seitz, W. (1940). Zur Frage des Extremitatencharakters der Tracheenkiemen von *Sialis flavilatera* L. *Z. Morph. Ökol. Tiere*, **37**, 214-275.

Selman, B. J. (1960a). On the tissue isolated in some of the larval appendages of *Sialis lutaria* L. at the larval-pupal moult. *J. Insect Physiol.*, **4**, 235-257.

Selman, B. J. (1960b). Tolerance of dehydration of the blood of *Sialis lutaria* L. *J. Insect Physiol.*, **6**, 81-83.

Selman, B. J. (1962). The fate of the blood cells during the life history of *Sialis lutaria* L. *J. Insect Physiol.*, **8**, 209-214.

Selman, B. J. (1965). The circulatory system of the alder fly *Sialis lutaria* L. *Proc. zool. Soc., Lond.*, **144**, 487-535.

Shaw, J. (1955a). The permeability and structure of the cuticle of the aquatic larva of *Sialis lutaria*. *J. exp. Biol.*, **32**, 330-352.

Shaw, J. (1955b). Ionic regulation and water balance in the aquatic larva *Sialis lutaria*. *J. exp. Biol.*, **32**, 353-382.

Staddon, B. W. (1955). The excretion and storage of ammonia by the aquatic larva of *Sialis lutaria*. *J. exp. Biol.*, **32**, 84-94.

Stein, F. (1838). Entwicklungs-Geschichte mehrerer Insectengattungen aus der Ordnung der Neuropteren. *Arch. Naturgesch.*, **4**, 315-333.

Sutcliffe, D. W. (1962). The composition of the haemolymph in aquatic insects. *J. exp. Biol.*, **39**, 325-343.

Sutcliffe, D. W. (1963). The chemical composition of haemolymph in insects and some other arthropods in relation to their phylogeny. *Comp. Biochem Physiol.*, **9**, 121-135.

Syms, E. E. (1935). Biological notes on British Megaloptera. *Proc. S. Lond. ent. nat. Hist. Soc.*, 1934-5, 121-124.

Tjeder, B. (1944). A note on the food of the adult *Sisyra fuscata* F. (Neuroptera, Sisyridae). *Ent. Tidskr.*, **65**, 203-204.

Vaňhara, J. (1970). The taxonomy and faunistics of the Czechoslovakian species of the order Megaloptera. *Acta ent. bohemoslov.*, **67**, 133-141.

Ward, P. H. (1965). A contribution to the knowledge of the biology of *Osmylus fulvicephalus*. *Entomologist's Gaz.* **16**, 175-182.

Wesenberg-Lund, C. (1917). *Furesøstudier. En Bathymetrisk botanisk zoologisk undersøgelse af Mølleaaens Søer.* Kobenhavn. A. F. Host. 208 pp.

Wesenberg-Lund, C. (1943). *Biologie der Susswasserinsekten.* Berlin. Springer, 682 pp.

Withycombe, C. L. (1923). Notes on the biology of some British Neuroptera (Planipennia). *Trans. R. ent. Soc. Lond.*, **1922**, 501-594.

Withycombe, C. L. (1925). Some aspects of the biology and morphology of the Neuroptera. With special reference to the immature stages and their possible phylogenetic significance. *Trans. R. ent. Soc. Lond.*, 1924, 303-411.

INDEX

Species names in parentheses are synonyms

PREFACE

The original publication on aquatic Megaloptera and Neuroptera was written by D. E. Kimmins and published by the Association in 1944 with a second edition in 1962. Some of the excellent illustrations of this earlier work are included in the present publication but more illustrations have been added and the text has been completely rewritten. As adults of species with terrestrial larvae may be confused with adults of species with aquatic larvae, a key to all families is included in the present work and some of the keys to species have been revised. The section on life cycles and ecology has been expanded and is now a more complete review of the literature.

Distribution maps have not been included because detailed records of the distribution of each species are not available. The list of references includes all the important publications and should therefore serve as a guide to the literature.

SBN 900386 27 4

ISSN 0367-1887

A KEY TO THE LARVAE AND ADULTS OF BRITISH FRESHWATER

Megaloptera

and

Neuroptera

with notes on their life cycles and ecology

by

J. M. ELLIOTT

Freshwater Biological Association

Illustrated by

D. E. KIMMINS and C. JOAN WORTHINGTON

1977

FRESHWATER BIOLOGICAL ASSOCIATION
SCIENTIFIC PUBLICATION No. 35